Mathematics through Paper Folding

ALTON T. OLSON

University of Alberta
Edmonton, Alberta

NATIONAL COUNCIL OF TEACHERS OF MATHEMATICS

Copyright © 1975 by
THE NATIONAL COUNCIL OF TEACHERS OF MATHEMATICS, INC.
1906 Association Drive, Reston, Virginia 22091

Second printing 1977

Library of Congress Cataloging in Publication Data:

Olson, Alton T
  Mathematics through paper folding.
  Bibliography: p.
  1. Mathematics—Study and teaching—Audio-visual
aids. 2. Paper work. I. Title.
QA19.P34038   516'.2   75-16115
ISBN 0-87353-076-4

Printed in the United States of America

Contents

Folding a straight line ▪ A straight line through a given point ▪ A line perpendicular to a given straight line ▪ The perpendicular to a line at a point on the line ▪ A line perpendicular to a given line and passing through a given point P not on the line ▪ The perpendicular bisector of a given line segment ▪ A line parallel to a given straight line ▪ A line through a given point and parallel to a given straight line ▪ The bisector of a given angle ▪ The location of equally spaced points along a line ▪ The formation of a right angle

Vertical angles ▪ The midpoint of the hypotenuse of a right angle ▪ The base angles of an isosceles triangle ▪ The intersection of the angle bisectors of a triangle ▪ The intersection of the perpendicular bisectors of the sides of a triangle ▪ The intersection of the medians of a triangle ▪ The area of a parallelogram ▪ The square on the hypotenuse is equal to the sum of the squares on the two other legs of a right triangle ▪ The diagonals of a parallelogram ▪ The median of a trapezoid ▪ The diagonals of a rhombus ▪ A line midway between the base and vertex of a triangle ▪ The sum of the angles of a triangle ▪ The area of a triangle ▪ The intersection of the altitudes of a triangle

The diameter of a circle ▪ The center of a circle ▪ The center of a circle of which only a portion (that includes the center) is available ▪ Equal chords and equal arcs in the same circle ▪ A diameter perpendicular to a chord ▪ A radius that bisects the angle between two radii ▪ Arcs of a circle intercepted by parallel lines ▪ The angle inscribed in a semicircle ▪ A tangent to a circle at a given point on the circle

Introduction

If mathematics educators and teachers had to choose the single most important principle for the learning of mathematics, they would probably allude to the importance of "active mathematical experiences." One intriguing way of adding an element of active experience to a mathematics class is to fold paper. Forming straight lines by folding creases in a piece of paper is an interesting way of discovering and demonstrating relationships among lines and angles. Once a relationship has been shown by folding paper, formal work on it later does not seem so foreign. Paper folding not only simplifies the learning of mathematics—it also builds an experiential base necessary for further learning.

The concepts and ideas of *motion,* or *transformation,* geometry are becoming standard fare for the mathematics curriculum. Paper folding offers many opportunities for illustrating these ideas. Folding a paper in half and making the halves coincident is an excellent physical model for a line reflection.

The exercises in this publication are appropriate at many different grade levels. Some exercises can profitably be done by students at a relatively advanced level—the entire section on conics, for example, is adapted for senior high school students. Other exercises, the simpler ones, have been enjoyed by elementary school pupils. Most of the introductory exercises would probably be appropriate for junior high school students. Many of the exercises are recreational and are of an enrichment nature. A few exercises are of a pattern type, such as the "dragon curves."

The only materials needed for paper-folding exercises are paper, felt pen, straightedge, and scissors. Although any type of paper may be used, waxed paper has a number of advantages: a crease becomes a distinct white line, and the transparency helps students "see" that in folding, lines and points are made coincident by placing one on the other.

Although paper folding is easy, it is not always easy to give clear instructions to students either orally or in writing. It is helpful to supplement demonstrations with directions and diagrams. In the text that follows, the diagrams are numbered with reference to the related exercise. They are not numbered consecutively. As the descriptions are read, the described folding should be performed. After these folding have been practiced, it is likely that the method can be extended to many more complex constructions.

1

In mathematics we always make certain basic assumptions on which we build a mathematical structure. In paper folding we assume the following postulates:

- Paper can be folded so that the crease formed is a straight line.
- Paper can be folded so that the crease passes through one or two given points.
- Paper can be folded so that a point can be made coincident with another point on the same sheet.
- Paper can be folded so that a point on the paper can be made coincident with a given line on the same sheet and the resulting crease made to pass through a second given point provided that the second point is not in the interior of a parabola that has the first point as focus and the given line as directrix. (A parabola forms the boundary between a convex region [interior] and a nonconvex region [exterior] of the plane.)
- Paper can be folded so that straight lines on the same sheet can be made coincident.
- Lines and angles are said to be congruent when they can be made to coincide by folding the paper.

If these assumptions are accepted, then it is possible to perform all the constructions of plane Euclidean geometry by folding and creasing.

Patterns for folding a great variety of polyhedra can be found in the following publications:

Cundy, H. M., and A. P. Rollett. *Mathematical Models.* 2d ed. London: Oxford University Press, 1961.

Hartley, Miles C. *Patterns of Polyhedrons.* Chicago: The Author, 1945. (No longer in print.)

Stewart, B. M. *Adventures among the Toroids.* Okemus, Mich.: The Author, 1970.

References on paper folding:

Barnett, I. A., "Geometrical Constructions Arising from Simple Algebraic Identities." *School Science and Mathematics* 38 (1938): 521–27.

Betts, Barbara B. "Cutting Stars and Regular Polygons for Decorations." *School Science and Mathematics* 50 (1950): 645–49.

Davis, Chandler, and Donald Knuth. "Number Representations and Dragon Curves—I." *Journal of Recreational Mathematics* 3 (April 1970): 66–81.

Joseph, Margaret. "Hexahexaflexagrams." *Mathematics Teacher* 44 (April 1951): 247–48.

Leeming, Joseph. *Fun with Paper*. Philadelphia: J. P. Lippincott Co., 1939.

Pedersen, Jean J. "Some Whimsical Geometry." *Mathematics Teacher* 65 (October 1972): 513–21.

Row, T. Sundara. *Geometric Exercises in Paper Folding*. Rev. ed. Edited by W. W. Beman and D. E. Smith. Gloucester, Mass.: Peter Smith, 1958.

Rupp, C. A. "On a Transformation by Paper Folding." *American Mathematical Monthly* 31 (November 1924): 432–35.

Saupe, Ethel. "Simple Paper Models of the Conic Sections." *Mathematics Teacher* 48 (January 1955): 42–44.

Uth, Carl. "Teaching Aid for Developing $(a + b)(a - b)$." *Mathematics Teacher* 48 (April 1955): 247–49.

Yates, Robert C. *Geometrical Tools*. St. Louis: Educational Publishers, 1949. (No longer in print.)

Since this publication is a revised edition of Donovan Johnson's classic *Paper Folding for the Mathematics Class*, a great deal of credit must go to him for providing so much of the inspiration and information that went into the making of this publication.

How to Fold the Basic Constructions

A variety of geometric figures and relationships can be demonstrated by using the following directions. If you have a supply of waxed paper and a couple of felt marking pens, you are all set for a new way of learning some mathematics.

1. Folding a straight line

Fold over any point P of one portion of a sheet of paper and hold it coincident with any point Q of the other portion. While these points are held tightly together by the thumb and forefinger of one hand, crease the fold with the thumb and forefinger of the other hand. Then extend the crease in both directions to form a straight line. From any point on the crease the distances to P and to Q are equal. Why must the crease form a straight line? (Fig. 1.)

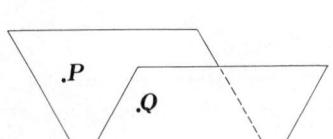

Fig. 1

Mathematically, the point P is called the *image* of point Q in a reflection in the line formed by the crease. Conversely, Q is the image of P in the same reflection.

2. A straight line through a given point

Carefully form a short crease that passes through the given point. Extend the crease as described previously. (Fig. 2.)

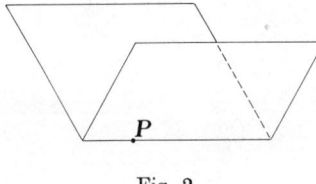

Fig. 2

3. A line perpendicular to a given straight line

Fold the sheet over so that a segment of the given line AB is folded over onto itself. Holding the lines together with the thumb and forefinger of one hand, form the crease as in exercise 1. (Fig. 3.)

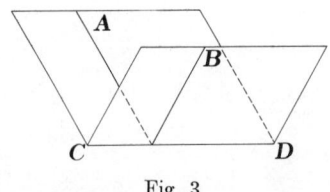

Fig. 3

4

The line AB is reflected onto itself by a reflection in the line formed by the crease. Why is the straight angle formed by the given line AB bisected by the crease CD?

4. The perpendicular to a line at a point on the line

Fold the paper so that a segment of the given line AB is folded over onto itself and so that the crease passes through the given point P. (Fig. 4.)

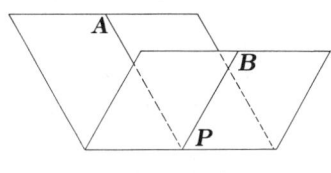

Fig. 4

Again the line AB is reflected onto itself in a reflection in the line formed by the crease. The point P is its own image in this reflection. Why is the fold through P perpendicular to AB?

5. A line perpendicular to a given line and passing through a given point P not on the line

Use the same method of folding as outlined in exercise 4. (Fig. 5.)

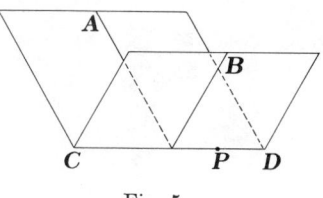

Fig. 5

6. The perpendicular bisector of a given line segment

Fold the paper so that the endpoints of the given line AB are coincident. Why is the crease CD the perpendicular bisector of AB? Locate any point on the perpendicular bisector. Is this point equally distant from A and B? (Fig. 6.)

What is the image of the line from a point on the perpendicular bisector to A when it is reflected in the perpendicular bisector?

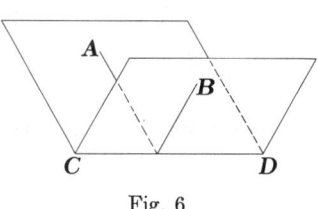

Fig. 6

7. A line parallel to a given straight line

First fold the perpendicular EF to the given line AB as in exercise 3. Next fold a perpendicular to EF. Why is this last crease CD parallel to the given line AB? (Fig. 7.)

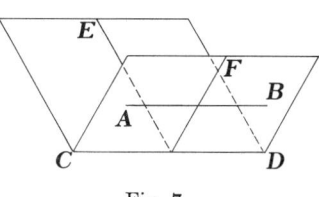

Fig. 7

5

8. A line through a given point and parallel to a given straight line

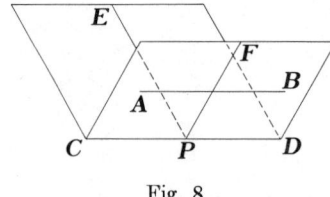

Fig. 8

First fold a line *EF* through the given point *P* perpendicular to the given line *AB* as in exercise 5. In a similar way, fold a line *CD* through the given point *P* and perpendicular to the crease *EF* formed by the first fold. Why does this crease provide the required line? (Fig. 8.)

9. The bisector of a given angle

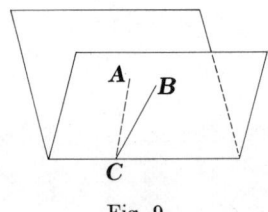

Fig. 9

Fold and crease the paper so that the legs *CA* and *CB* of the given angle *ACB* coincide. Why must the crease pass through the vertex of the angle? How can you show that the angle is bisected? (Fig. 9.) An angle is reflected onto itself in a reflection in its bisector.

10. The location of equally spaced points along a line

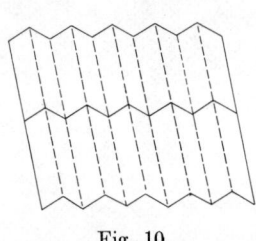

Fig. 10

Establish any convenient length as the unit length by folding a segment of the line onto itself. Form several equal and parallel folds by folding back and forth and creasing to form folds similar to those of an accordion. (Fig. 10.)

11. The formation of a right angle

Any of the previous constructions involving perpendiculars can be used to produce right angles. See exercises 3, 4, 5, and 6.

Geometric Concepts Related to Reflections Illustrated by Paper Folding

12. Vertical angles

AB and CD are two lines that intersect at O. Fold and crease the paper through vertex O, placing BO on CO. Do AO and DO coincide? Are vertical angles congruent? (Fig. 12A.)

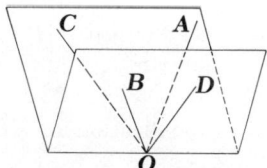

Fig. 12A

With different-colored felt pens draw the lines AB and CD intersecting at O. For convenience, make each of a pair of vertical angles less than 45°. Fold two creases EF and GH in the paper so that they are perpendicular at O. Neither of these creases should be in the interiors of the vertical angles. (Fig. 12B.)

Fold the paper along line EF. Follow this by folding along line GH. Now the vertical angles should coincide. Line AB should coincide with itself, and line CD should coincide with itself. What differences do you see between the results of figures 12A and 12B?

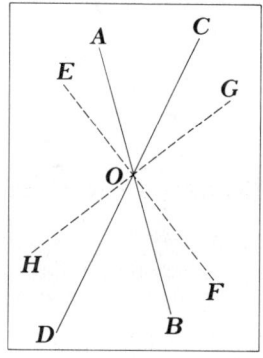

Fig. 12B

Mathematically, one of the vertical angles in figure 12B has been rotated 180° with O as the center. Also, in the pair of vertical angles, one of the angles is the image of the other after a reflection in O.

13. The midpoint of the hypotenuse of a right triangle

a) Draw any right triangle ABC (fig. 13).

b) Find the midpoint D of hypotenuse AB by folding. Fold the line from the midpoint D to C.

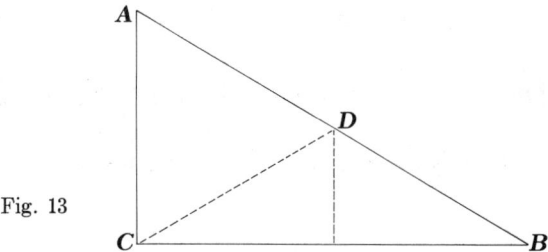

Fig. 13

c) Compare *CD* and *BD* by folding the angle bisector of *BDC*. What is the image of *CD* in a reflection in this angle bisector?

14. The base angles of an isosceles triangle

The isosceles triangle *ABC* is given with *AB* congruent to *BC*. Fold line *BD* perpendicular to *AC*. Compare ∠*A* and ∠*C* by folding along *BD*. (Fig. 14.)

The image of ∠*A* is ∠*C* in a reflection in *BD*. What is the image of ∠*C*? Are angles *A* and *C* congruent?

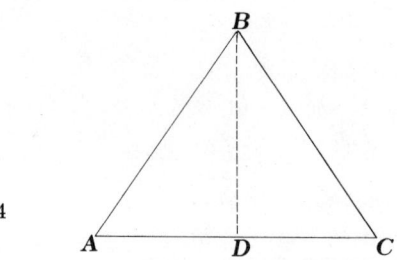

Fig. 14

15. The intersection of the angle bisectors of a triangle

Fold the bisectors of each angle of the given triangle. Do the bisectors intersect in a common point? What is the point of intersection of the angle bisectors called? Fold the perpendiculars from this point of intersection to each of the sides of the triangle. Compare *ID*, *IE*, and *IF* by folding. (Fig. 15.)

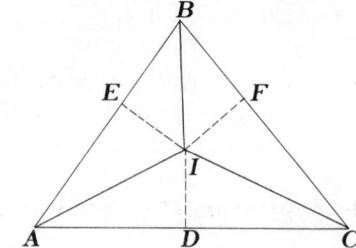

Fig. 15

ID is the image of *IF* in a reflection in *IC*. What is the image of *IE* in a reflection in *IB*? What conclusions can be made about *ID*, *IE*, and *IF*?

8

16. The intersection of the perpendicular bisectors of the sides of a triangle

Fold the perpendicular bisectors of each side of the given acute triangle. What is the common point of intersection of these lines called? Fold lines from this point to each vertex of the triangle. Compare these lengths by folding. (Fig. 16.)

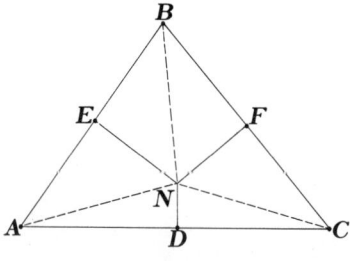

Fig. 16

AN is the image of CN in a reflection in ND. What is the image of NB in a reflection in NE? What conclusions can be made about AN, BN, and NC?

17. The intersection of the medians of a triangle

Bisect the three sides of the given triangle. Fold lines from the midpoint of each side to the opposite vertex. What is the point of intersection of these lines called? Try balancing the triangle by placing it on a pin at the intersection of the medians. What is this point called?

Fold a line perpendicular to BE through G. E' is the point on the median that coincides with E when the triangle is folded along this perpendicular line. E' is the image of E in a reflection in the line perpendicular to BE through G. If another line perpendicular to BE is folded through E', then what is the image of B in a reflection in this line? (Fig. 17.)

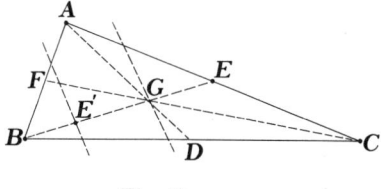

Fig. 17

Repeat this same procedure for the other two medians. What can be concluded about the position of G on each of the three medians?

9

18. The area of a parallelogram

Cut out a trapezoid with one side CB perpendicular to the parallel sides. Fold the altitude DE. Fold CF parallel to AD. For convenience the trapezoid should be cut so that the length of EF is greater than the length of FB. Fold FG perpendicular to AB. After folding triangle FBC over line FG, make another fold at HJ so that B coincides with E and C coincides with D. Does F coincide with A? Are triangles ADE and FCB congruent? (See fig. 18. Figure 18 and others so noted are included in Appendix C, where they appear large enough for ditto masters to be made from them.)

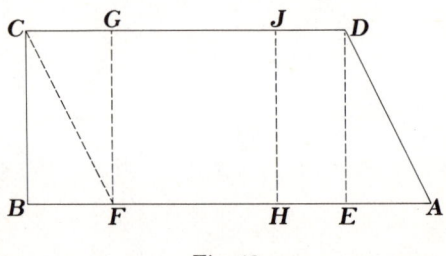

Fig. 18

Mathematically, the result of reflecting triangle FCB in FG and then in HJ is a *slide,* or *translation,* in the direction of B to E. Why is this terminology appropriate?

When triangle FCB is folded back, $ADCF$ is a parallelogram. When triangle ADE is folded back, $DCBE$ is a rectangle. Is rectangle $BCDE$ equal in area to parallelogram $ADCF$? What is the formula for the area of a parallelogram?

19. The square on the hypotenuse is equal to the sum of the squares on the two other legs of a right triangle

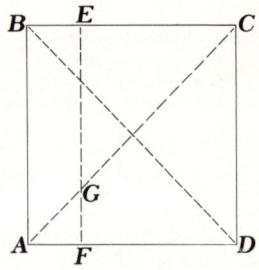

Fig. 19A

Use a given square $ABCD$. Make a crease EF perpendicular to sides AD and BC. Fold diagonals AC and BD. (See fig. 19A. Also, see Appendix C for·an enlarged model of figure 19A.) Fold along diagonal AC. Crease the resulting double thickness along GF and GE (fig. 19B). When the square is opened flat, the line HI will have been formed (fig. 19C). HI is the image of FE in a reflection in AC.

10

Folding along the diagonal BD will form the lines JK and LM.

Fold lines EK, KM, MH, and HE. Let the measure of $EK = c$, $EC = a$, and $CK = b$. Then equate the area of $EKMH$ to the sum of the areas of $NOPG$ and the four triangles ENK, OKM, MPH, and HGE. If this equation is written in terms of a, b, and c, then what is the result?

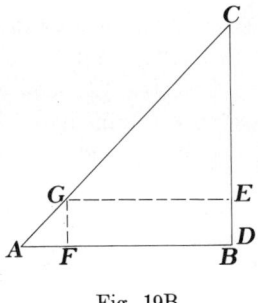

Fig. 19B

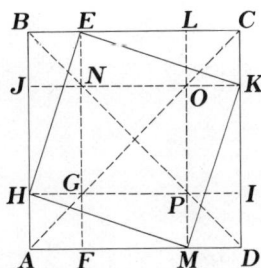

Fig. 19C

20. The diagonals of a parallelogram

Fold the diagonals of a given parallelogram. Compare the lengths of BE and AE by folding the bisector of angle BEA. Are the diagonals equal in length? Fold a line perpendicular to BD through E. Compare the lengths of EB and ED by folding along this perpendicular line. What is the image of D in a reflection in this perpendicular line? Repeat the same procedure for the other diagonal AC. Do the diagonals of a parallelogram bisect each other? (See fig. 20. Also, see Appendix C for an enlarged model of figure 20.)

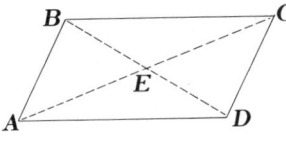

Fig. 20

21. The median of a trapezoid

Fold the altitudes at both ends of the shorter base of the trapezoid $ABCD$. Bisect each nonparallel side and connect these midpoints with a crease EF. Compare DG and CH with GI and HJ respectively by folding along EF. What are the images of DG and CH in a reflection in EF?

11

What is the image of CD in this same reflection? Fold lines perpendicular to AB through E and F. What are the images of A and B in reflections in these respective perpendicular lines? How does the sum of CD and AB compare with the median EF? (See fig. 21. Also, see Appendix C for an enlarged model of figure 21.)

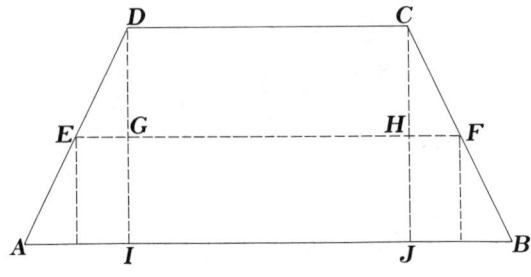

Fig. 21

22. The diagonals of a rhombus

Fold the diagonals of a given rhombus $ABCD$. Compare AO and BO to OC and OD respectively by folding along the diagonals. What is the image of AO in a reflection in BD? What is the image of angle ABD in a reflection in BD? What conclusions can you make about the diagonals of a rhombus? Is triangle ABD congruent to triangle CBD? (See fig. 22. Also, see Appendix C for an enlarged model of figure 22.)

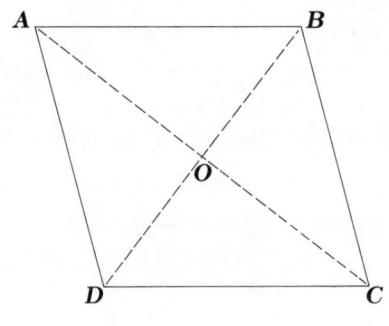

Fig. 22

23. A line midway between the base and vertex of a triangle

Bisect two sides of the triangle ABC (fig. 23). Fold a line EF through the midpoints. Fold the altitude to the side that is not bisected. Compare BG and GD by folding along line EF. What is the image of BG in a reflection in EF? Bisect GD. Fold a line perpendicular to BD through H.

12

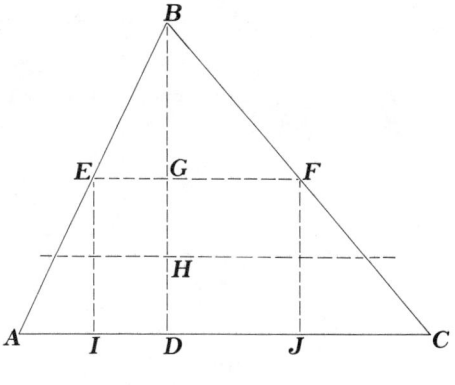

Fig. 23

What is the image of *EF* in a reflection in this perpendicular line? Is *EF* parallel to *AC*? Fold lines perpendicular to *AC* through *E* and through *F*. What are the images of *A* and of *C* when reflected in *EI* and *FJ* respectively? How does the length of *EF* compare with the length of *AC*?

24. The sum of the angles of a triangle

a) Fold the altitude *BD* of the given triangle *ABC* (fig. 24A).

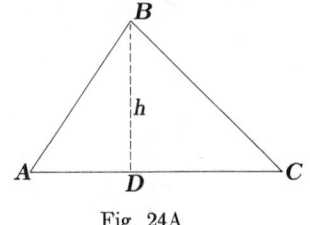

Fig. 24A

b) Fold the vertex *B* onto the base of the altitude, *D* (fig. 24B). How is line *EF* related to line *AC*? How are *AE* and *EB* related?

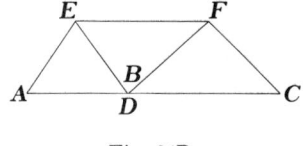

Fig. 24B

c) Fold the base angle vertices *A* and *C* to the base of the altitude, *D* (fig. 24C). Does the sum of ∠*A*, ∠*B*, and ∠*C* make up a straight angle?

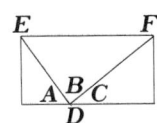

Fig. 24C

13

25. The area of a triangle

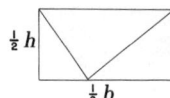

$\frac{1}{2}h$ $\frac{1}{2}b$

Fig. 25

In figure 24C, the rectangular shape has sides whose measures are equal to one-half the base AC of triangle ABC and one-half the altitude BD (fig. 25). What is the area of the rectangle? How are the areas of this rectangle and the original triangle related? What is the area of the triangle?

26. The intersection of the altitudes of a triangle

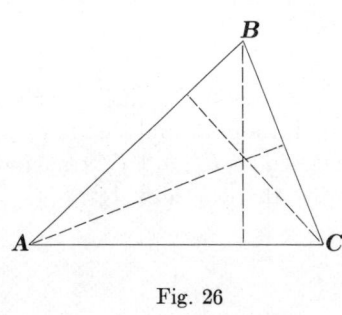

Fig. 26

Fold the altitudes to each side of the given triangle (fig. 26). Do they intersect in a common point? What is the intersection point of the altitudes called? Are there any relationships among the distances from the point of intersection of the altitudes to the vertices and bases of the triangle? Repeat this exercise for an obtuse triangle.

14

Circle Relationships Shown by Paper Folding

27. The diameter of a circle

Fold the circle onto itself (fig. 27).
Does the fold line AB bisect the circle?
What name is given to line AB? What is
the image of the circle when it is reflected
in line AB? The circle is said to have line
symmetry with respect to line AB.

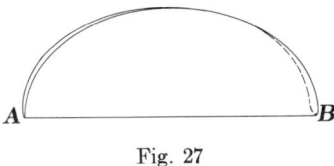

Fig. 27

28. The center of a circle

Fold two mutually perpendicular diameters (fig. 28). Are the diameters
bisected? At what point do the diameters intersect? What is the image of
AO in a reflection in CD?

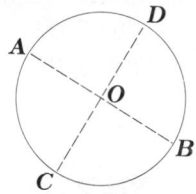

Fig. 28

29. The center of a circle of which only a portion (which includes the center) is available

Fold a chord AB and a chord BC (fig. 29). Fold the perpendicular bisector of AB. From any point on this perpendicular bisector, the distance
to A is the same as the distance to B. How could this be shown? Fold the
perpendicular bisector of BC. It intersects the other perpendicular bisector
at M. What is true of AM, MB, and MC? Why is M the center of the
circle?

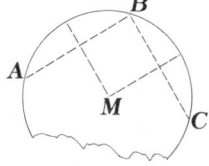

Fig. 29

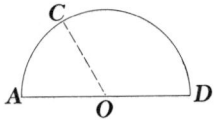

Fig. 30A

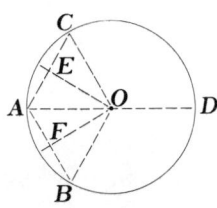

Fig. 30B

30. Equal chords and equal arcs in the same circle

Locate the center O of the circle by folding two diameters. Fold the circle along a diameter AD. From some point C, fold the semicircle along CO (fig. 30A). This forms two radii, CO and BO (fig. 30B). How does arc AC compare with arc AB? What is the image of arc AC in a reflection in AD? Fold chords AB and AC. How does chord AC compare with chord AB? How does central angle COA compare with central angle AOB? Fold lines through O perpendicular to AC and to AB. By folding, compare AE with EC and AF with FB. What is the image of EC in a reflection in EO? Answer the same question for a reflection in AD. Compare EO with FO by folding along AD. What generalizations can be made about equal chords and equal arcs of the same circle?

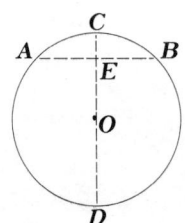

Fig. 31

31. A diameter perpendicular to a chord

Fold any chord AB (fig. 31). Fold a diameter CD perpendicular to this chord. Compare the segments AE and EB of the given chord. Compare the subtended arcs AC and CB.

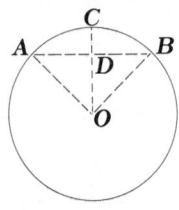

Fig. 32

32. A radius that bisects the angle between two radii

Fold any two radii, AO and BO (fig. 32). Fold the chord AB. Fold the bisector OC of the angle between the radii AO and BO. How is the bisector of angle AOB related to the chord AB? What is the image of arc AC in a reflection in angle bisector CO?

16

33. Arcs of a circle intercepted by parallel lines

Fold any diameter AB of circle O (fig. 33). Fold two chords, each perpendicular to AB. What are the images of E and F in a reflection in AB? Compare arc EF to arc CD by folding.

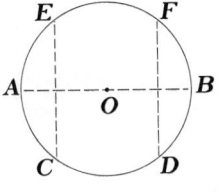

Fig. 33

34. The angle inscribed in a semicircle

Fold diameter AB (fig. 34). Fold a chord AC. Extend AC. Likewise, fold CB and extend it. What is the image of CB is a reflection in AC? What is the size of the angle formed by the chords AC and BC?

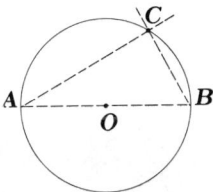

Fig. 34

35. A tangent to a circle at a given point on the circle

Fold the diameter of the given circle passing through the given point P on the circle (fig. 35). At P, fold the line perpendicular to the diameter. Why is this perpendicular line tangent to the circle? If this perpendicular line passed through another point Q on the circle, then what would be true of the image of Q in a reflection in the diameter?

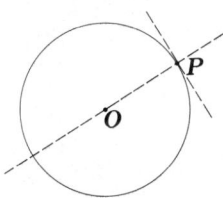

Fig. 35

17

Algebra by Paper Folding

36. (ax + by) · (cx + dy)

a) Let any rectangular sheet of paper represent a rectangle with dimensions x and $x + y$ (fig. 36A).

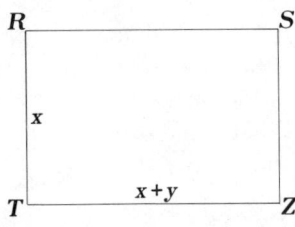

Fig. 36A

b) To determine y, fold the upper left-hand vertex down to the bottom edge (fig. 36B). Fold along VU. The measures of RT and UZ are x and y respectively. Fold Z to point W on UV. Fold along WL. (Fig. 36C.)

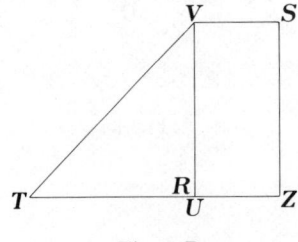
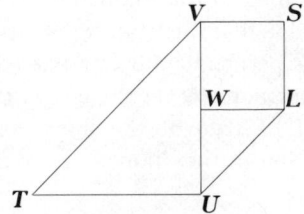

Fig. 36B   Fig. 36C

c) Unfold and return to the original rectangle. $RTVU$ is a square x units on each side. $UVSZ$ is a rectangle with dimensions x and y. $UWLZ$ is a square y units on each side. (Fig. 36D.)

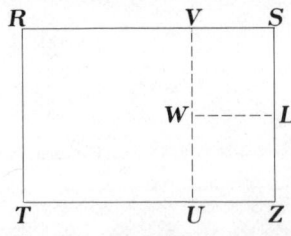

Fig. 36D

18

d) Cut out several model rectangles with sides x and y and several squares with sides of x and of y. These will be needed in the following exercises. For convenience, color one face of the model rectangles red, blue, or some other bright color, and leave the opposite face white.

e) Label the rectangle and squares as in figure 36E. The square formed by M, N, N, and Q is $x + y$ on a side. Its area is $(x + y) \cdot (x + y)$. Since the areas of M, N, and Q are x^2, $x \cdot y$, and y^2, respectively, we have $(x + y)(x + y) = x^2 + xy + xy + y^2 = x^2 + 2xy + y^2$.

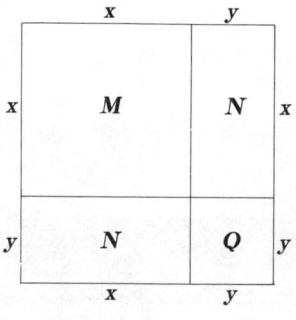

Fig. 36E

f) Mathematically, the area of the rectangle in figure 36F is $(2x + 3y) \cdot (2x + y)$. Summing the areas of the Ms, Ns, and Qs, respectively, we obtain $(2x + 3y)(2x + y) = 4x^2 + 8xy + 3y^2$.

x	x	y	y	y
M	M	N	N	N
M	M	N	N	N
N	N	Q	Q	Q

Fig. 36F

g) Assume that the product $(3x - 2y)(2x - y)$ is to be found. Arrange the various rectangles and squares so that they make up a rectangle that is $3x + 2y$ on one side and $2x + y$ on an adjacent side. To begin with, all the rectangles should be white side up. To represent $3x - 2y$, turn rectangles 4, 5, 9, and 10 and squares 14 and 15 over,

19

exposing the colored side. To represent $2x - y$, turn rectangles 11, 12, and 13 and squares 14 and 15 over in the same manner. Now squares 14 and 15 have been turned over twice, again exposing the white sides. (Fig. 36G.)

	x	x	x	-y	-y
x	1	2	3	4	5
x	6	7	8	9	10
-y	11	12	13	14	15

Fig. 36G

The squares 1, 2, 3, 6, 7, 8, 14, and 15 represent positive products. The rectangles 4, 5, 9, 10, 11, 12, and 13 each represent the product $-x \cdot y$. Thus, $(3x - 2y)(2x - y) = 6x^2 - 7xy + 2y^2$.

h) Assume that the product $(x + y)(x - y)$ is to be found. In a manner similar to that of the preceding exercise, arrange the squares and rectangles in such a way that they make up a square that is $x + y$ on a side. All the rectangles and squares should be white side up. To represent $x - y$, turn rectangle 2 and square 4 over. Since rectangles 2 and 3 represent products of different sign, $(x + y)(x - y) = x^2 - xy + xy - y^2 = x^2 - y^2$. (Fig. 36H.)

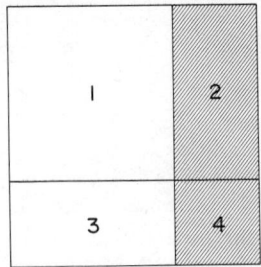

Fig. 36H

20

37. Multiplication and division of *a* and *b*

Fold two perpendicular lines, $X'X$ and $Y'Y$, intersecting at O. Fold a series of equally spaced points on the two lines. Be sure to include O in the points. These folded points will form a coordinate system for the plane of the paper.

Let OU be $+1$. Define OA and OB as directed line segments representing a and b respectively (fig. 37A). Join U to A by folding a line through these two points. Through B fold a line parallel to AU and let P be the point of intersection of this line and $X'X$. Now OP represents the product of a and b in magnitude and sign. In figure 37A, a was positive and b was negative.

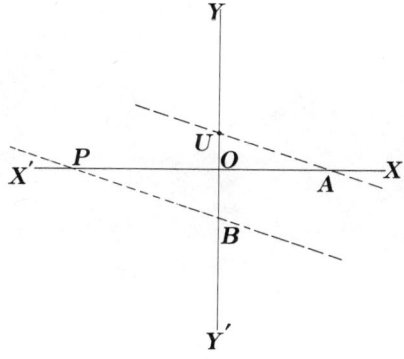

Fig. 37A

Fold a line passing through A and B. Fold a line passing through U parallel to AB. Let Q be the point of intersection of this line and $X'X$. Then OQ represents the quotient a/b in magnitude and sign (fig. 37B).

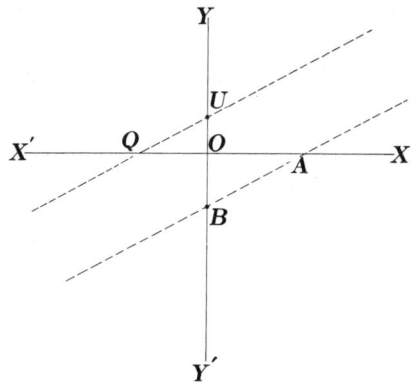

Fig. 37B

21

38. Solving $x^2 - px + q = 0$, p and q integers

Fold two intersecting lines, $X'X$ and $Y'Y$, intersecting at 0. Coordinatize each of the lines by folding equally spaced points. Let OP and OQ represent p and q respectively. Fold perpendiculars to $X'X$ and $Y'Y$ at P and Q, intersecting at M. Fold a line determined by M and U. OU is the line representing $+1$. Now find the midpoint of UM by folding. Let T be this midpoint. Now U is reflected in some line that passes through T so that the image of U is on $X'X$. There will be two such points if $x^2 - px + q = 0$ has two real, unequal roots. If these two points are R and S, then OR and OS represent the roots in both magnitude and sign. (Fig. 38.)

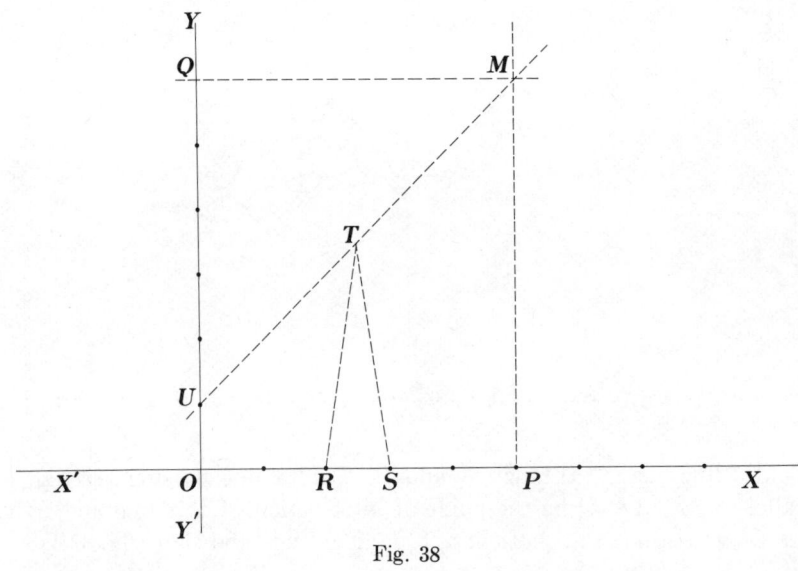

Fig. 38

To find R and S, fold the paper, without creasing, along lines that pass through T. By adjusting the fold, it is possible to make U coincide with $X'X$ at R and S. The procedure is illustrated below using the equation $x^2 - 5x + 6 = 0$. Notice that $OR = 2$ and $OS = 3$ in measure.

A circle can be drawn through Q, U, R, and S. How can you show this? Why must OR and OS be representations of the roots of the equation?

Star and Polygon Construction

39. Triangle

Fold any three nonparallel creases that will intersect on the sheet (fig. 39).

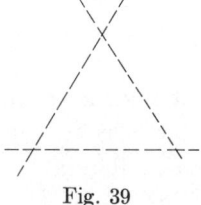

Fig. 39

40. Regular hexagon, equilateral triangle, and three-pointed star

Fold and crease a piece of paper. This crease is shown as AB in figure 40A. From some point O on AB, fold OB to position OB' so that angle $AOB' \cong$ angle $B'OE$. The congruent angles are most easily obtained by means of a protractor. They can also be approximated by judicious folding. Crease OB so that OA falls on OE (fig. 40B.). In figure 40B, XZ is perpendicular to OE, and the measures of OX and OW are equal. Cutting along XW results in a regular hexagon. An equilateral triangle results when a cut is made along XZ. Cutting along XY results in a three-pointed star.

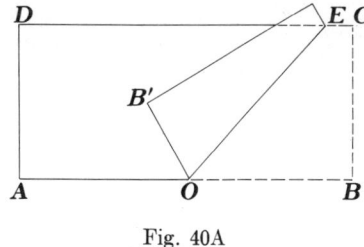 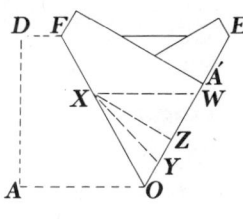

Fig. 40A   Fig. 40B

41. Equilateral triangle

a) Fold the median EF of rectangle $ABCD$ (fig. 41).

b) Fold vertex A onto EF so that the resulting crease, GB, passes through B. Denote by J the position of A on EF. Return to original position by unfolding. Fold line GJ, extending it to H.

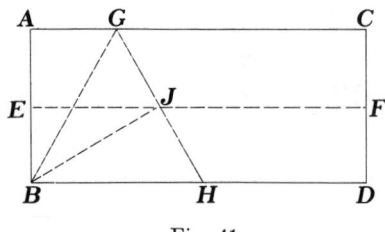

Fig. 41

c) By folding, show that *BJ* is perpendicular to *GH*.

d) What is the image of angle *GBJ* in a reflection in *BJ*? What is the image of angle *ABG* in a reflection in *BG*?

e) Fold the angle bisector of angle *BGH* and of angle *GHB*. What conclusions can be made after reflections in these angle bisectors?

f) Why is triangle *BGH* an equilateral triangle?

42. Isosceles triangle

Fold the perpendicular bisector of side *AB* of rectangle *ABCD*. From any point *P* on the perpendicular bisector, fold lines to vertices *A* and *B*. What conclusions can be made after a reflection in this perpendicular? (Fig. 42.)

Why is triangle *ABP* an isosceles triangle?

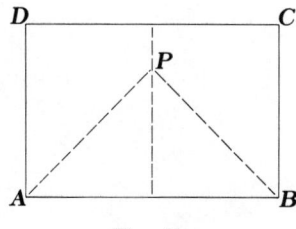

Fig. 42

43. Hexagon

Fold the three vertices of an equilateral triangle to its center (fig. 43). How is this center found?

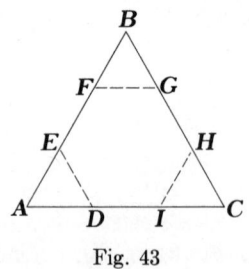

Fig. 43

24

Is the hexagon *DEFGHI* equilateral? How does the area of triangle *ABC* compare with that of hexagon *DEFGHI*?

44. Regular octagon, square, and four-pointed star

Fold a piece of paper in half and crease. Call the resulting line *AB*. Fold the perpendicular bisector of *AB*. Call this *OE*. (Fig. 44.) Fold *OA* and *OB* over so that they coincide with *OE* and crease *OF*. Mark point *W* so that triangle *OXW* is isosceles, and mark point *Z* so that *XZ* is perpendicular to *OF*. Cutting along *XW* will result in a regular octagon. A square results from a cut along *XZ*. Cutting along *XY* gives a four-pointed star.

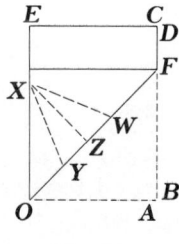

Fig. 44

45. Rectangle

Fold any straight line *AB*. At points *D* and *F* on *AB*, fold lines perpendicular to *AB*. At point *G* on line *CD*, fold a line perpendicular to *CD*. This perpendicular line intersects *EF* at *H*. (Fig. 45.) Show by folding that *GH* is perpendicular to *EF*. What is the image of *EF* in a reflection in *GH*?

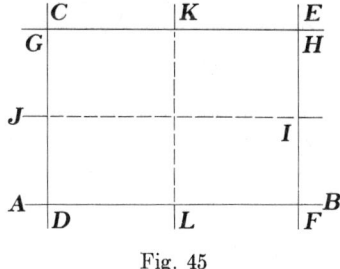

Fig. 45

Bisect side *EF* by folding. Fold a line perpendicular to *HF* through midpoint *I*. By reflecting rectangle *DFHG* in the line *JI*, what relationships among lines and angles appear to be true? Fold a line perpendicular to *GH* through midpoint *K*. Reflect the rectangle *DFHG* in *KL* and note what relationships appear to be true.

25

46. Square

Fold a rectangle so that one of the right angles is bisected (line *BE*). Fold *FE* perpendicular to *AD* (fig. 46). Why is *ABFE* a square? What is the image of *F* in a reflection in *BE*? What is the image of *C* in a reflection in *BD*?

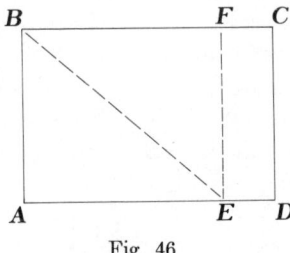

Fig. 46

47. Other relationships in the square derived by reflections

Find the midpoints of the sides of *ABCD* by folding. Fold the diagonals *AC* and *BD*. Fold all possible lines determined by midpoints *E*, *F*, *G*, and *H*. (See fig. 47A. Also, see Appendix C for an enlarged model of figure 47A.)

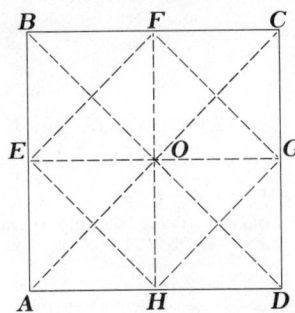

Fig. 47A

a) What are the images of *B*, *F*, and *C* in a reflection in *EG*? From this result, what line segments are congruent?

b) What is the image of angle *BOF* in a reflection in *EG*? Consequently, what angles are congruent?

c) What is the image of angle *FOC* in a reflection in *AC*? In *EG*? In *BD*? In *FH*?

d) What is the image of *C* in a reflection in *FG*?

e) What lines can be shown to be perpendicular by folding?

f) How does the area of inscribed square *EFGH* compare with the original square *ABCD*?

26

If the area of the original square $ABCD$ is 1 square foot, what are the areas of the other squares formed by folding the corners to the center? (Fig. 47B.)

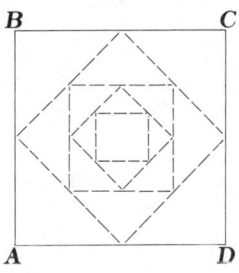

Fig. 47B

48. Octagon

Fold a square $ABCD$ to obtain the midpoints E, F, G, and H. Fold the inscribed square $EFGH$. By folding, bisect the angles formed by the sides of the original square and the sides of the inscribed square $EFGH$ (fig. 48). Fold the various diagonals AC, BD, EG, and FH. What are the images of HI, EI, EJ, and JF in a reflection in line FH? What are the images of the sides of octagon $EJFKGLHI$ in reflections in lines AC, BD, and EG? Why is $EJFKGLHI$ a regular octagon?

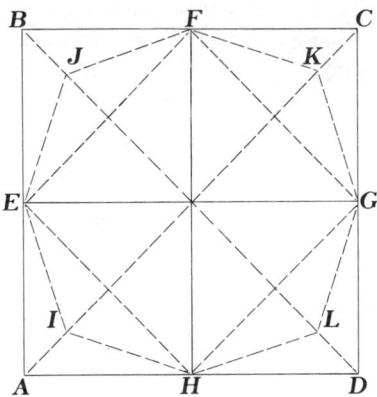

Fig. 48

27

49. Regular decagon, regular pentagon, and five-pointed star

Fold a piece of paper in half and crease. Call this line AB. If O is the midpoint of AB, fold and crease along line OE so that angle AOB equals one-half of angle BOE in measure (fig. 49A). This angle relationship can be assured by using a protractor or can be approximated by careful folding. Fold OE over so that it coincides with OB. Crease line OF (fig. 49B). Crease along OE so that OA falls along OF (fig. 49C). Triangle OXW is an isosceles triangle. Triangle OXZ is a right triangle. Cutting along XW results in a regular decagon. Cutting along XZ results in a regular pentagon. A five-pointed star is produced when a cut is made along XY.

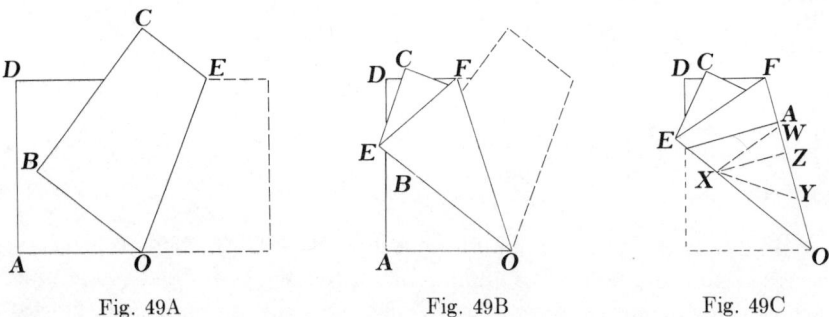

Fig. 49A   Fig. 49B   Fig. 49C

50. Six-pointed star, regular hexagon, and regular dodecagon

Fold a piece of paper in half. Call this line AB. Fold A over on B and crease along OE. Fold A and B over and crease along OF so that angle EOA equals angle AOF in measure (fig. 50A). This angle congruence can be assured by using a protractor or can be approximated by careful folding. Crease on OA, folding OF over to fall along OE (fig. 50B). Triangle OXW is isosceles. Triangle OXZ is a right triangle. Cutting along XW, XZ, and XY respectively will result in a regular dodecagon, regular hexagon, and a six-pointed star. Interesting snowflake patterns can be made by cutting notches in the six-pointed star design.

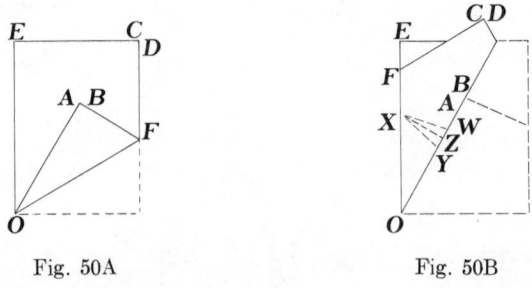

Fig. 50A   Fig. 50B

Polygons Constructed by Tying Paper Knots

51. Square

Use two strips of paper of the same width.

a) Fold each strip over onto itself to form a loop and crease. Why are the angles that are formed right angles? (Fig. 51A.)

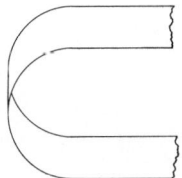

Fig. 51A

b) Insert an end of one strip into the loop of the other so that the strips interlock. Pull the strips together tightly and cut off the surplus. Why is the resulting polygon a square? (Fig. 51B.)

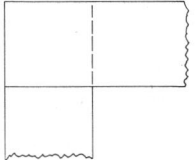

Fig. 51B

52. Pentagon

Use a long strip of constant width. Tie an overhand knot (fig. 52A). Tighten the knot and crease flat (fig. 52B). Cut the surplus lengths. Unfold and consider the set of trapezoids formed by the creases. How many trapezoids are formed? Compare the trapezoids by folding. What conclusions can be made about the pentagon obtained?

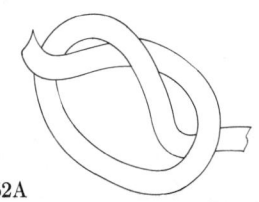

Fig. 52A

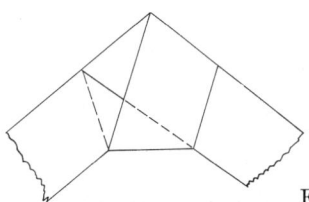

Fig. 52B

53. Hexagon

Use two long strips of paper of equal width. Tie a square knot as shown in figure 53A. Tighten and crease it flat to produce a hexagon. It may be easier to untie the knot and fold each piece separately according to figure 53B. After tightening and flattening, cut off the surplus lengths. Unfold and consider the trapezoids formed. How many trapezoids are formed on each strip? Compare the sizes of these trapezoids.

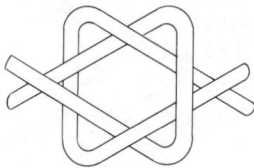

Fig. 53A

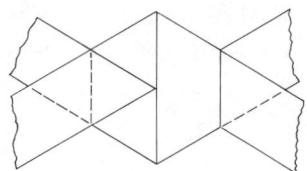

Fig. 53B

54. Heptagon

Use a long strip of constant width. Tie a knot as illustrated in figure 54A. Tighten and crease flat (fig. 54B). How many trapezoids are formed when the knot is untied?

Fig. 54A

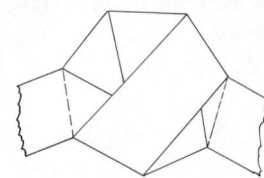

Fig. 54B

55. Octagon

Use two long strips of the same width. First, tie a loose overhand knot with one strip like that for the pentagon above. Figure 55 shows this tie with the shaded strip going from 1-2-3-4-5. With the second strip, start at 6, pass over 1-2 and under 3-4. Bend up at 7. Pass under 4-5 and 1-2. Bend up at 8. Pass under 3-4 and 6-7. Bend up at 9. Pass over 3-4, under 7-8 and 4-5, emerging at 10. Tighten and crease flat. Cut surplus lengths 1, 5, 6, and 10 (fig. 55).

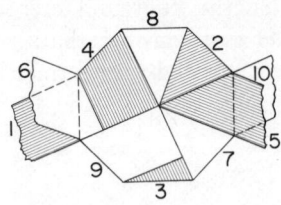

Fig. 55

This construction is not easy. Another tack might be to analyze the knots and their trapezoids to determine the lengths and the sizes of angles involved. Using a protractor, a ruler, and the obtained information would make the constructions considerably easier.

Symmetry

56. Line symmetry

Fold a line in a sheet of paper. Cut out a kite-shaped figure similar to figures 56A and 56B. Fold this figure along any other line. What differences do you note between the foldings in the two lines? The first fold is a symmetry line for the figure. What is the image of the figure in a reflection in the first fold line?

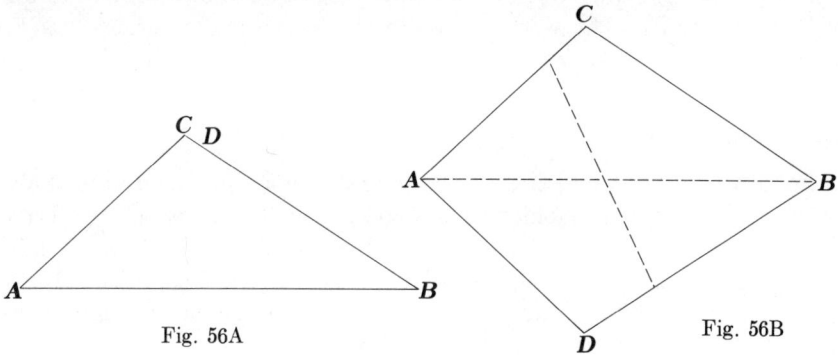

Fig. 56A   Fig. 56B

57. Line and point symmetry

Fold two perpendicular creases. Keeping the paper folded, cut out a plane curve with a scissors (fig. 57A).

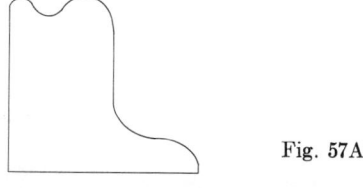

Fig. 57A

What are the images of the figure when they are reflected in *AB* and in *CD*? Line *EF* is drawn so that it passes through *O* and is different from *AB* and *CD*. (Fig. 57B.)

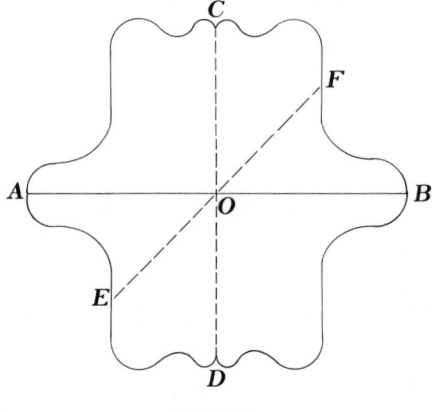

Fig. 57B

Is *EF* a line of symmetry for the figure? How can you show this? How is *O* related to *EF*? Answer these questions for various positions of *EF*. Point *O* is a point of symmetry for the figure. Can you see why?

58. Symmetrical design

Fold two perpendicular creases, dividing the paper into quadrants. Fold once more, bisecting the folded right angles. Keep the paper folded. Trim the edge opposite the 45° angle so that all folded parts are equal. While the paper remains folded, cut odd-shaped notches and holes. Be sure to leave parts of the edges intact. (Fig. 58A.) When the paper is unfolded, a symmetrical design is apparent (fig. 58B).

Fig. 58A

Fig. 58B

32

Conic Sections

59. Parabola

Draw any straight line m as a directrix. Mark a point F not on the given line as the focus. Fold a line perpendicular to line m. Mark the point of intersection of line m and the line perpendicular to m. Call it point G. Fold the paper over so that point F coincides with point G and crease. Call the point of intersection of this crease and the perpendicular line H. (Fig. 59.) Repeat this operation twenty to thirty times by using different lines perpendicular to m. The point H will be on a parabola with focus F and directrix m. The creases formed by folding point F onto point G are tangents to the parabola. The tangents are said to "envelop" the parabolic curve.

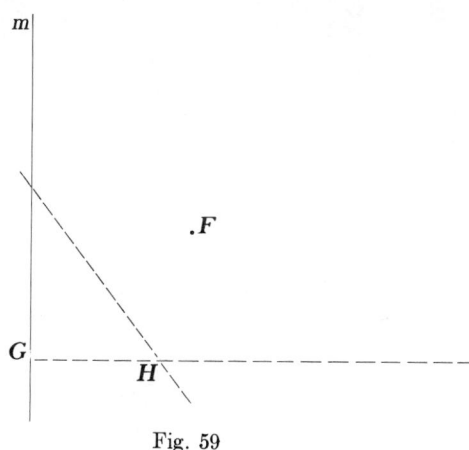

Fig. 59

What is the image of FH when reflected in the crease formed by the coincidence of F and G? What geometric facts concerning tangents to parabola can be obtained from this?

Imagine that the inside of the parabolic curve is a mirrored surface. Rays of light, which are parallel to the lines perpendicular to m, strike the mirror. Where are these rays of light reflected after striking the mirror?

60. Ellipse

Draw a circle with center O. Locate a point F inside the circle. Mark a point X on the circle. Fold the point F onto X and crease. Fold the diameter that passes through X. The point of intersection of this diameter and

33

the crease is called P. Repeat this procedure twenty to thirty times by choosing different locations for X along the circle. Each crease is tangent to an ellipse with foci F and O. (Fig. 60.) What is the image of PX under a reflection in ZY? Show how the measure of FP plus the measure of PO is equal to a constant. Thus, P is on the ellipse, with O and F as foci. Imagine that ZY is a mirror. Why would a ray of light passing through F and P be reflected through O? Let R be any point along ZY other than P. Show that the sum of the measures of FR and RO is greater than the sum of the measures of FP and PO.

Repeat this experiment by using various locations for F. What effect does this have on the resulting ellipses?

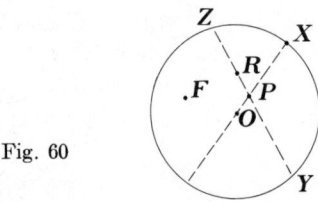

Fig. 60

61. Hyperbola

Draw a circle with center O. Locate a point F outside the circle. Mark a point X on the circle. Fold F onto X and crease. This crease is tangent to a hyperbola with O and F as foci. Fold a diameter through X. The point of intersection of the diameter and the crease is called P. (Fig. 61.)

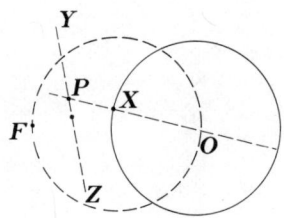

Fig. 61

What is the image of FP in a reflection in YZ? Show that the measure of FP minus the measure of PO equals a constant. Thus, point P is on the hyperbola with foci F and O. Repeat this procedure twenty to thirty times by choosing different locations for X along the circle.

Draw a circle that has OF as a diameter. Include the points of intersection of the two circles as choices for the location of X. The resulting creases are asymptotes for the hyperbola. What is the image of the hyperbola in a reflection in OF? What is the image of the hyperbola in a reflection in a line perpendicular to OF at the midpoint of OF?

34

62. Similarity and enlargement transformations

a) Draw a triangle ABC. Mark a point D outside the triangle. Fold line AD. Fold point D onto A and crease. The point of intersection of this crease and line AD is called A'. Repeat the same procedure for points B and C in order to locate points B' and C' (fig. 62A).

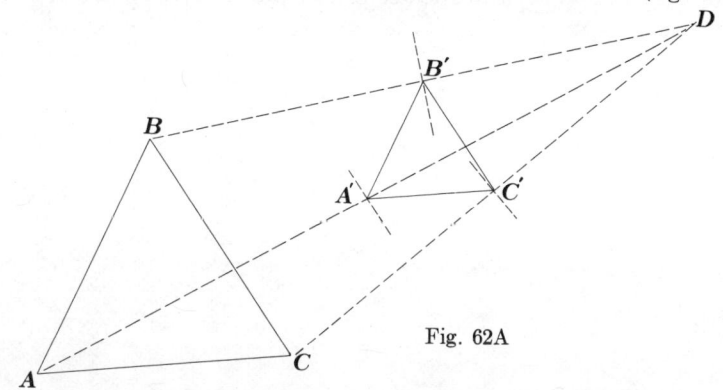

Fig. 62A

How is triangle ABC related to triangle $A'B'C'$? How do the areas of these two triangles compare?

b) Draw a triangle ABC and point D outside this triangle. Reflect point D in a line perpendicular to AD at point A. Call this image point A'. Repeat the same procedure with points B and C in order to locate points B' and C'. Do the same with point X. Where is the image point X'? (Fig. 62B). How does triangle $A'B'C'$ compare with triangle ABC?

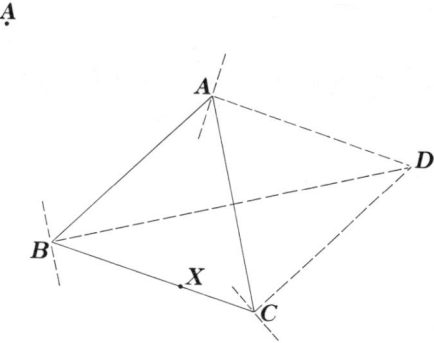

Fig. 62B $\dot{C}'$

35

c) Draw a triangle ABC and points D and E outside this triangle. Use the procedure from (a) with point D to locate triangle $A'B'C'$. Repeat this procedure with triangle $A'B'C'$ and point E to locate triangle $A''B''C''$ (fig. 62C). How is triangle $A''B''C''$ related to triangle ABC? How do their areas compare? Fold lines AA'', BB'', and CC''. What conclusions can be made after making these folds?

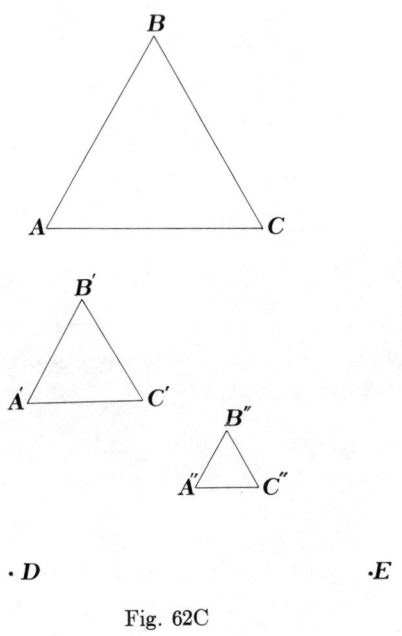

Fig. 62C

Recreations

63. Möbius strip

Use a strip of paper at least 1½ inches wide and 24 inches long. To make a Möbius strip, give one end a half-turn (180°) before gluing it to the other end (fig. 63). If you draw an unbroken pencil mark on the strip, you will return to the starting point without crossing an edge. Thus, this strip of paper has only one surface. Stick the point of a scissors into the center of the paper and cut all the way around. You will be surprised by the result! Cut the resulting band down the center for a different result. After two cuts how many separate bands do you have?

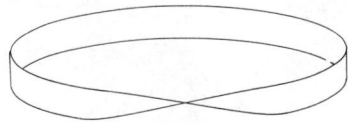

Fig. 63

64. Hexaflexagon

The hexaflexagon requires a paper strip that is at least six times its width in length.

a) First fold the strip to locate the center line CD at one end of the strip (fig. 64A).

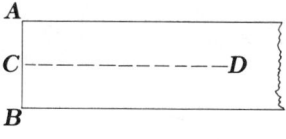

Fig. 64A

b) Fold the strip so that B falls on CD and the resulting crease AE passes through A (fig. 64B). Where would the image of A be in a reflection in BE? What kind of a triangle is ABE?

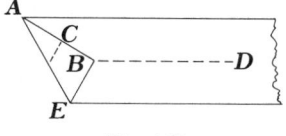

Fig. 64B

c) Fold the strip back so that the crease EG forms along BE (fig. 64C). What kind of a triangle is EGA? Next fold forward along GA, forming another triangle. Continue folding back and forth until ten equilateral triangles have been formed. Cut off the excess of the strip as well as the first right triangle ABE.

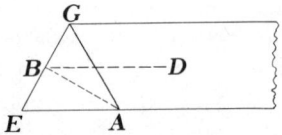

Fig. 64C

d) Lay the strip in the position shown in figure 64D and number the triangles accordingly.

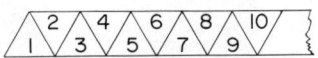

Fig. 64D. Front

e) Turn the strip over and number as in figure 64E. Be sure that triangle 11 is behind triangle 1.

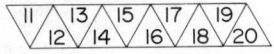

Fig. 64E. Back

f) To fold the hexaflexagon, hold the strip in the position shown in figure 64D. Fold triangle 1 over triangle 2. Then fold triangle 15 onto triangle 14 and triangle 8 onto triangle 7. Insert the end of the strip, triangle 10, between triangles 1 and 2. If the folding now gives the arrangements shown in figures 64F and 64G, glue triangle 1 to 10. If not, recheck the directions given.

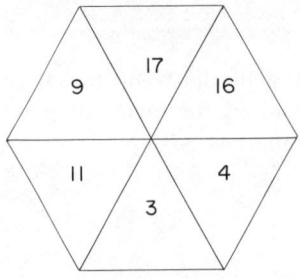

Fig. 64F

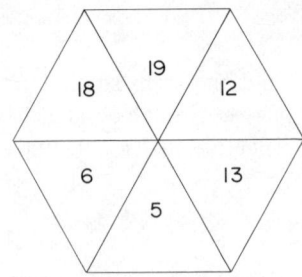

Fig. 64G

38

The hexagon can be folded and opened to give a number of designs. Two of these designs are given in figures 64F and 64G. The designs open easily by folding in the three single edges, thus forming a three-cornered star and opening out the center. How many different designs can be obtained?

65. Approximating a 60° angle

Cut a strip of paper two inches wide and about twenty inches long. Cut one end of the strip off and label the line of cutting t_0. By folding, bisect the angle formed by t_0 and the edge of the strip. Label the bisector t_1 and the two congruent angles formed x_0. The line t_1 intersects the other edge of the strip at A_1. By folding, bisect the obtuse angle formed at A_1 by t_1 and the edge of the strip. This procedure is continued until the lengths of t_k and t_{k+1} appear to be congruent and the angles X_k and X_{k+1} appear to be congruent. These angles X_k approach 60° in measure. (Fig. 65.) It is surprising that no matter what angle X_0 is used in the beginning, angles X_k always approach 60° in measure.

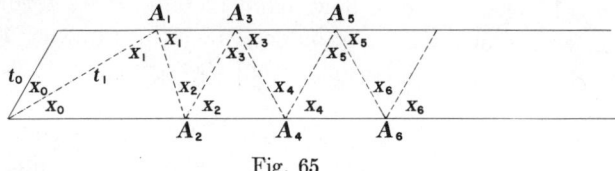

Fig. 65

66. Trisecting an angle

An interesting variation on exercise 65 takes place on a piece of paper whose straight edges are not parallel (fig. 66). In this situation, angle X_k approaches $\theta/3$ in measure. Thus, we have a way of approximating the trisection of angle θ. For convenience, choose A_0 as far away from B as possible. Also, to assure a convenient convergence, choose t_0 so that X_0 is approximately $\theta/3$ in measure.

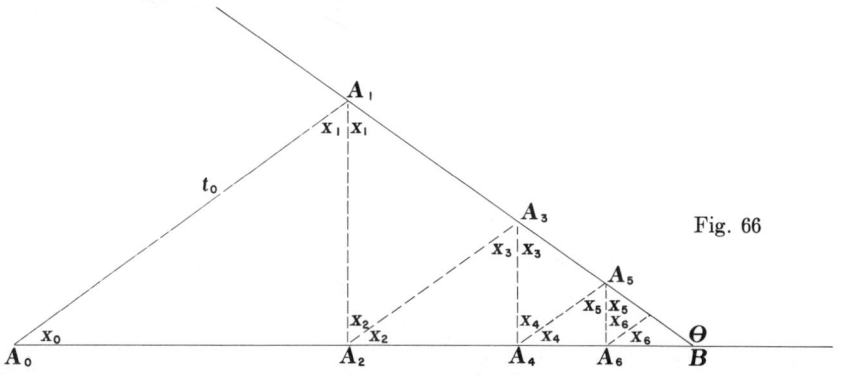

Fig. 66

67. Dragon curves

Take a long strip of paper and fold it in half from right to left. When it is opened, it has one crease, which points downward (fig. 67A). Fold the paper in half two times from right to left. When it is opened, it has three creases. Reading from left to right, the first two point downward and the third points upward (fig. 67B). For three folding-in-half operations, the pattern of creases is (left to right) $DDUDDUU$, where D and U represent creases that point downward and upward respectively.

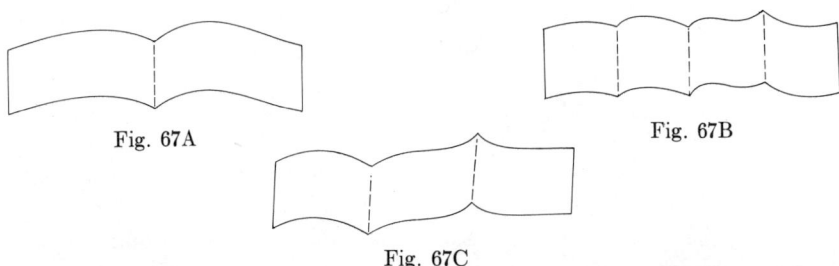

Fig. 67A   Fig. 67B

Fig. 67C

After n folding operations, how many rectangles are formed and how many creases are formed? Can you determine the sequence of Ds and Us for four folding-in-half operations from the sequences that result from the first three foldings?

Modify the folding above by alternately folding the ends from left to right and then from right to left. The formulas for determining the number of areas and the number of creases formed after n foldings will not change, but the sequence of Ds and Us used in describing the creases does change. Can you figure out how to predict the pattern for $n + 1$ folds, knowing the pattern for n folds?

Another interesting modification is to use a trisecting fold rather than a bisecting fold. Fold the strip so that the pattern after one trisection fold is DU (fig. 67C).

How many areas and how many creases are formed after n trisection-folding operations? Can you determine the sequence of Ds and Us for four trisection foldings, knowing the sequence for three trisection foldings?

68. Proof of the fallacy that every triangle is isosceles

Fold the bisector of the vertex angle and the perpendicular bisector of the base (fig. 68). These creases will intersect outside the triangle, which contradicts the assumption that these lines meet inside the triangle.

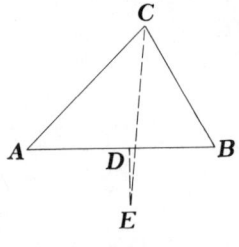

Fig. 68

69. Cube

a) Fold a piece of paper down to form a square and remove the excess strip. The edge of the cube that will eventually be formed will be one-fourth the side of this square (fig. 69A).

b) Fold the paper from corner to corner and across the center one way through the midpoint of the sides (fig. 69B). The fold across the center should be in the opposite direction to that of the corner-to-corner folds.

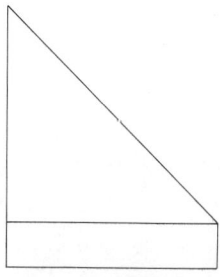

Fig. 69A

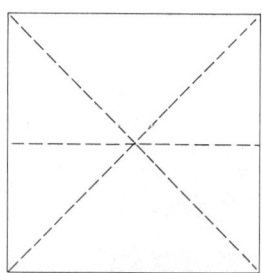

Fig. 69B

c) Let the paper fold naturally into the shape shown in figure 69C.

d) Fold the front *A* and *B* down to point *C* (fig. 69D).

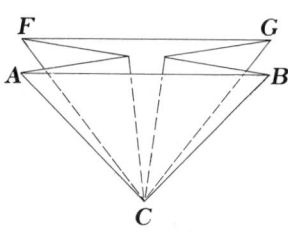

Fig. 69C

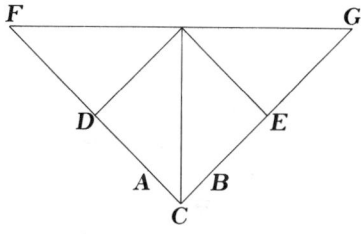

Fig. 69D

e) Turn it over and do the same for the back corners *F* and *G*. A smaller square will result (fig. 69E).

f) The corners on the sides *D* and *E* are now double. Fold the corners *D* and *E* so that they meet in the center. Turn the square over and do the same for the corners on the back side (fig. 69F).

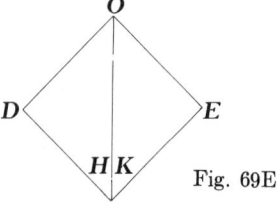

Fig. 69E

Fig. 69F

g) One end of figure 69F will now be free of loose corners. Fold the loose corners on the opposite end, *H* and *K*, outward on the front to form figure 69G. Do the same for the corresponding corners on the back.

h) Fold points *H* and *K* inward to the center. Do the same with the points on the back of the form (fig. 69H).

Fig. 69G

Fig. 69H

i) Open folds *D* and *E* and tuck triangles *LHM* and *KNP* into the pockets in *D* and *E*. Do the same with the points on the back (fig. 69I).

j) Blow sharply into the small hold found at *O* and the cube will inflate. Crease the edges and the cube is finished (fig. 69J).

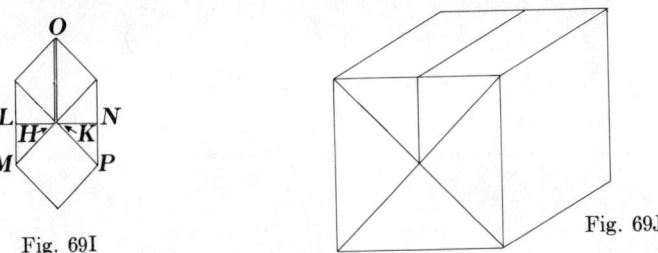

Fig. 69I

Fig. 69J

70. A model of a sphere

Cut three equal circles out of heavy paper. Cut along the lines as shown in figures 70A, 70B, and 70C. Bend the sides of figure 70A toward each

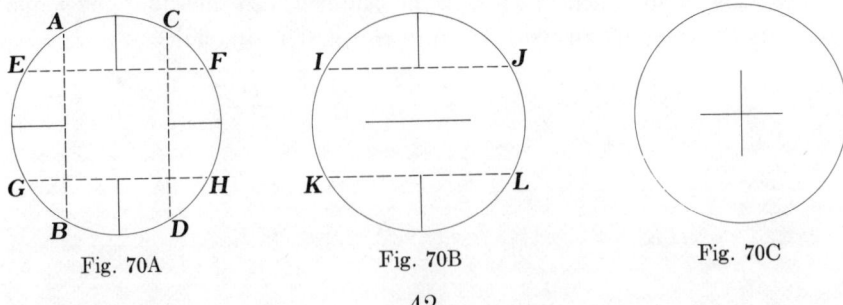

Fig. 70A

Fig. 70B

Fig. 70C

42

other along the dotted lines *AB* and *CD* and pass this piece through the cut in the center of figure 70B. Open figure 70A after it has been pushed through figure 70B.

Bend the sides of figure 70A along the dotted lines *EF* and *GH* and bend figure 70B along the dotted lines *IJ* and *KL*. Pass figures 70A and 70B through the cross-shaped cut in figure 70C. This will form the sphere model shown in figure 70D. This model is suitable for demonstrating latitude and longitude, time zones, and spherical triangles. It can also be used as a geometric Christmas tree decoration or in a mobile. If the model is to be made out of cardboard, figures 70A and 70C should be cut into two semicircles and fitted into figure 70B.

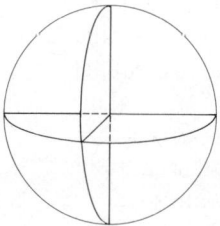

Fig. 70D

71. Pop-up dodecahedron

Cut two patterns as shown in figure 71A out of cardboard. Fold lightly along the dotted lines. Place these patterns together as shown in figure 71B and attach with a rubber band. Toss the model into the air and it will form a dodecahedron. If the first attempt is not successful, change the rubber band or use a different type of cardboard.

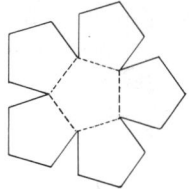

Fig. 71A

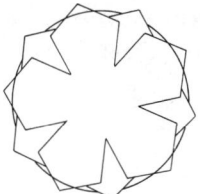

Fig. 71B

72. Patterns for polyhedra

Cut the following patterns from cardboard. Fold along the dotted lines. Use the tabs for gluing. (See Appendix C for enlarged models of figures 72A–G.)

43

Stellated polydedra can be made by attaching pyramids to each face of these regular polyhedra. Each pyramid should have a base congruent to the face of the polyhedron.

A less frustrating alternative to the "tab and glue" method of constructing polyhedra is the "cardboard and rubber band" method. To use this method, cut out each face of a polyhedron separately. On each edge of these pieces, cut a narrow tab, notched at each end and folded back. Fasten the pieces together along matching tabs secured by rubber bands. Stretch the rubber bands along the tabs and secure them in the notches. Tabs one-fourth inch in width seem to be best for securing the rubber bands.

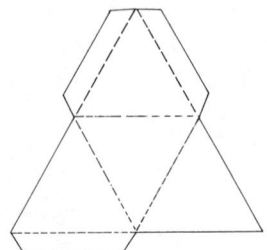

Fig. 72A. Tetrahedron

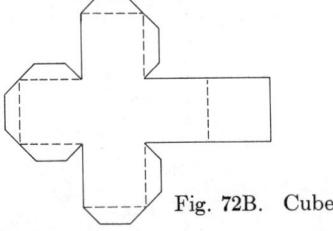

Fig. 72B. Cube

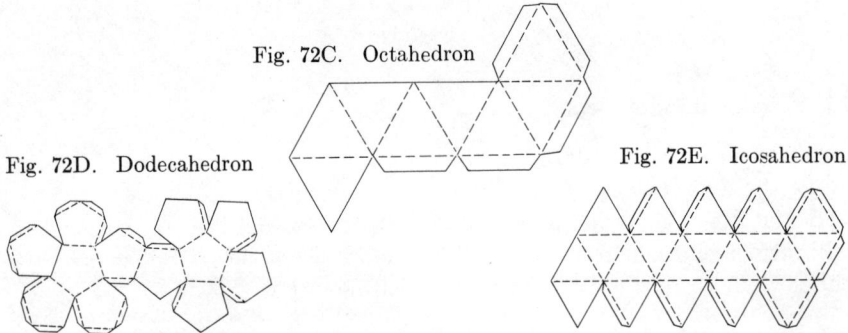

Fig. 72C. Octahedron

Fig. 72D. Dodecahedron

Fig. 72E. Icosahedron

Different polyhedra can be made by experimenting with regular polygons of three, four, five, and six sides. Obviously, all these polygons must have edges that are of equal length (figs. 72F and 72G).

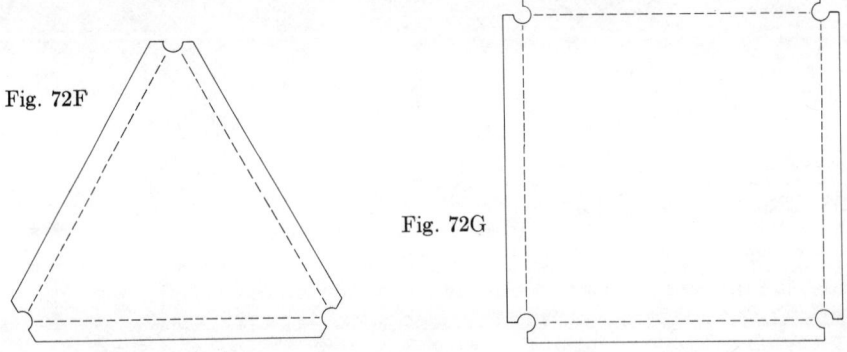

Fig. 72F

Fig. 72G

44

Appendix A
Plane Geometry Theorems and Related Exercises

In the following listing, certain theorems from plane geometry are given. After each theorem, related exercises from this monograph are noted.

1. In a plane, through a given point on a given line, there is one and only one line perpendicular to the given line. (Exercise 4)

2. In a plane, there is one and only one line perpendicular to a given line through a given point not on the line. (Exercise 5)

3. A segment has one and only one midpoint. (Exercise 6)

4. An angle has one and only one bisector. (Exercise 9)

5. Vertical angles are congruent. (Exercise 12)

6. The measure of the median to the hypotenuse of a right triangle is equal in measure to half the hypotenuse. (Exercise 13)

7. If two sides of a triangle are congruent, then the angles opposite these sides are congruent. (Exercise 14)

8. The three lines that bisect the angles of triangle ABC are concurrent at a point I that is equidistant from the lines AB, BC, and AC. (Exercise 15)

9. The three lines that are in the plane of triangle ABC and are the perpendicular bisectors of the sides of the triangle are concurrent at a point that is equidistant from the vertices A, B, and C. (Exercise 16)

10. The three medians of a triangle are concurrent at a point whose distance from any one of the vertices is two-thirds the length of the median from that vertex. (Exercise 17)

11. The area of a parallelogram is the product of the measures of a base and the altitude to that base. (Exercise 18)

12. The square of the measure of the hypotenuse of a right triangle is equal to the sum of the squares of the measures of the other two sides. (Exercise 19)

13. The diagonals of a parallelogram bisect each other. (Exercise 20)

14. The segment that joins the midpoints of the nonparallel sides of a trapezoid is parallel to the bases, and its measure is one-half the sum of the measures of the bases. (Exercise 21)

15. The diagonals of a rhombus are perpendicular to each other. (Exercise 22)

16. A diagonal of a rhombus bisects the angles formed at the related vertices. (Exercise 22)

17. The segment joining the midpoints of two sides of a triangle is parallel to the third side and is equal to one-half of its measure. (Exercise 23)

18. The sum of the measures of the angles of a triangle is 180°. (Exercise 24)

19. The area of any triangle is equal to one-half the product of the measures of any one of its bases and the altitude to that base. (Exercise 25)

20. The three altitude lines of a triangle are concurrent at a point. (Exercise 26)

21. In a circle, the minor arcs of congruent chords are congruent. (Exercise 30)

22. A diameter that is perpendicular to a chord bisects that chord. (Exercise 31)

23. In a circle, congruent central angles intercept congruent minor arcs. (Exercise 32)

24. If two parallel lines intersect a circle, then the intercepted arcs are congruent. (Exercise 33)

25. An angle inscribed in a semicircle is a right angle. (Exercise 34)

26. A tangent to a circle is perpendicular to the radius drawn to the point of contact. (Exercise 35)

27. If two angles of one triangle are congruent respectively to two angles of another triangle, then the triangles are similar. (Exercise 37)

Appendix B
Some Additional Theorems That Can Be Demonstrated by Paper Folding

1. The median from the vertex of the angle included by the congruent sides of an isosceles triangle bisects that angle.

2. The median from the vertex of the angle included by the congruent sides of an isosceles triangle is perpendicular to the third side.

3. The bisector of the angle included by the congruent sides of an isosceles triangle bisects the side opposite that angle.

4. Any two medians of an equilateral triangle are congruent.

5. If two distinct coplanar lines are intersected by a transversal that makes a pair of alternate interior angles congruent, the lines are parallel.

6. If two distinct coplanar lines are intersected by a transversal that makes a pair of corresponding angles congruent, the lines are parallel.

Appendix C
Large-Scale Figures

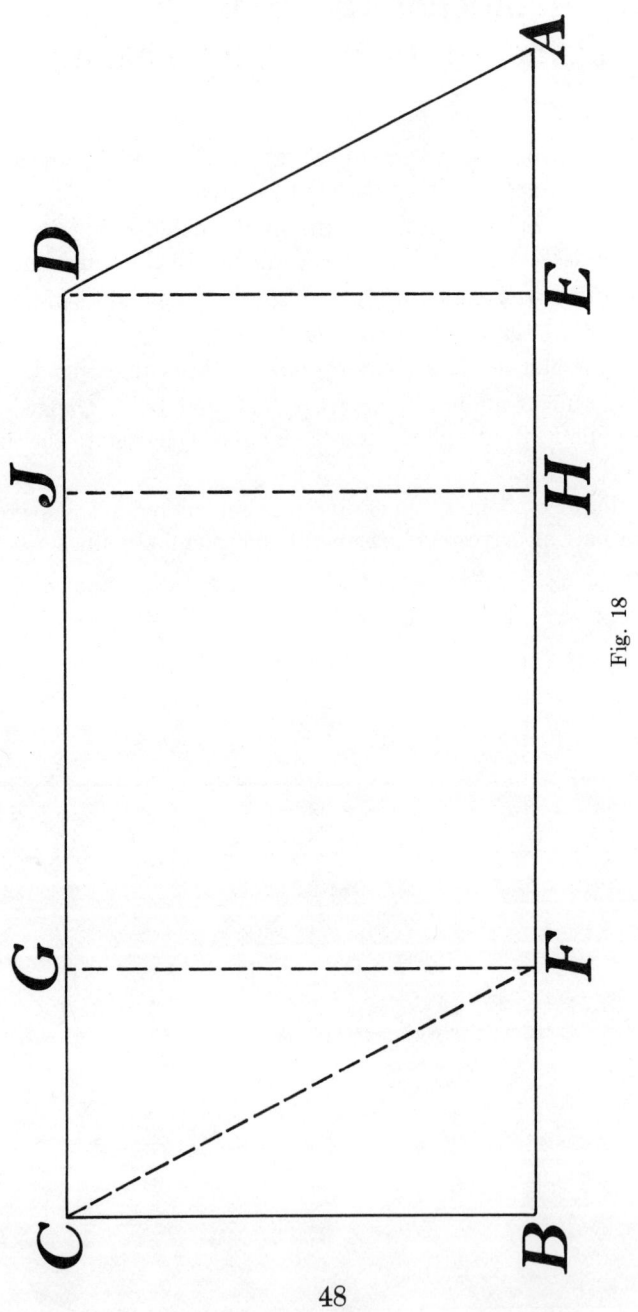

Fig. 18

48

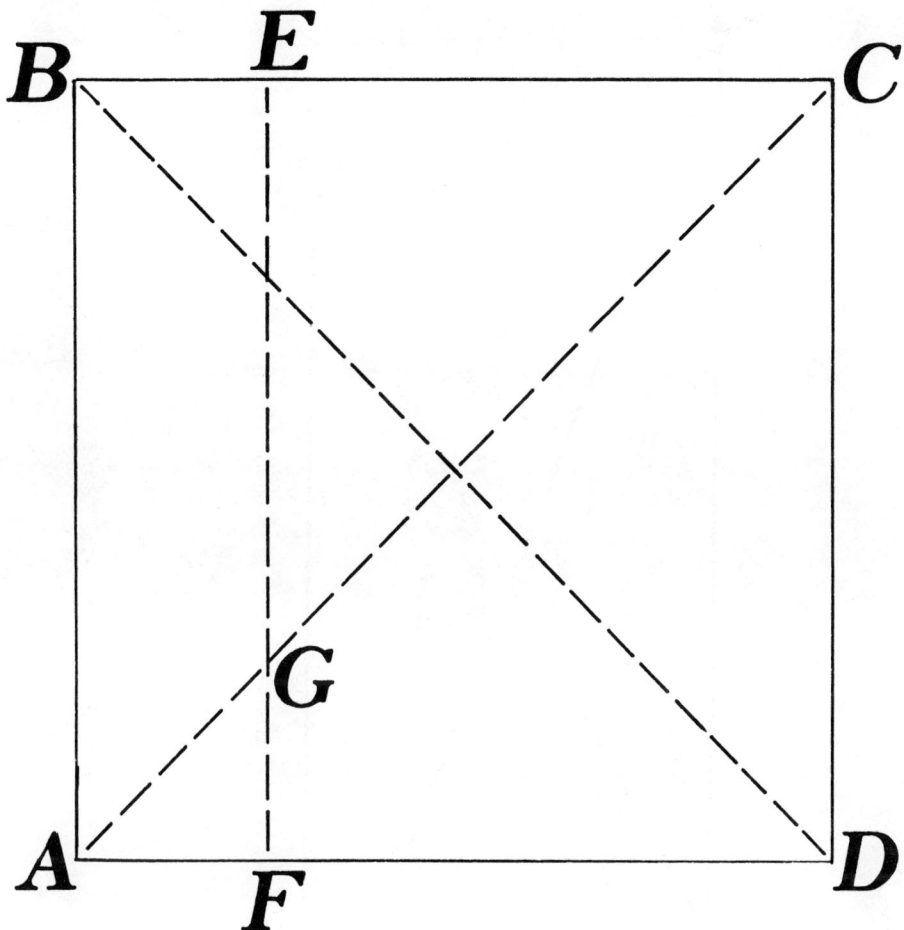

Fig. 19A

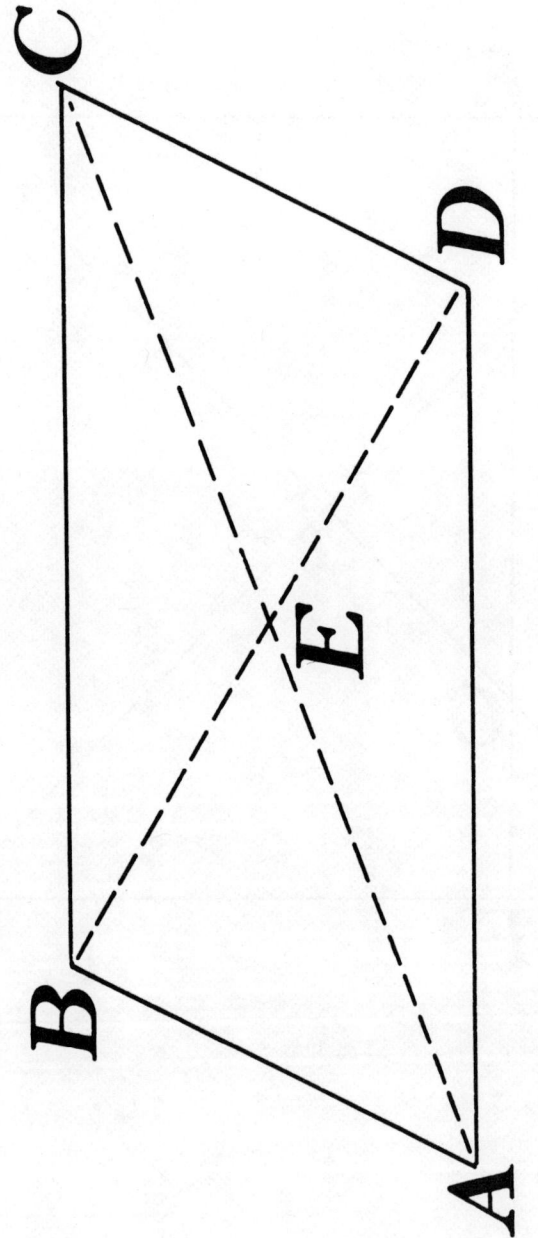

Fig. 20

50

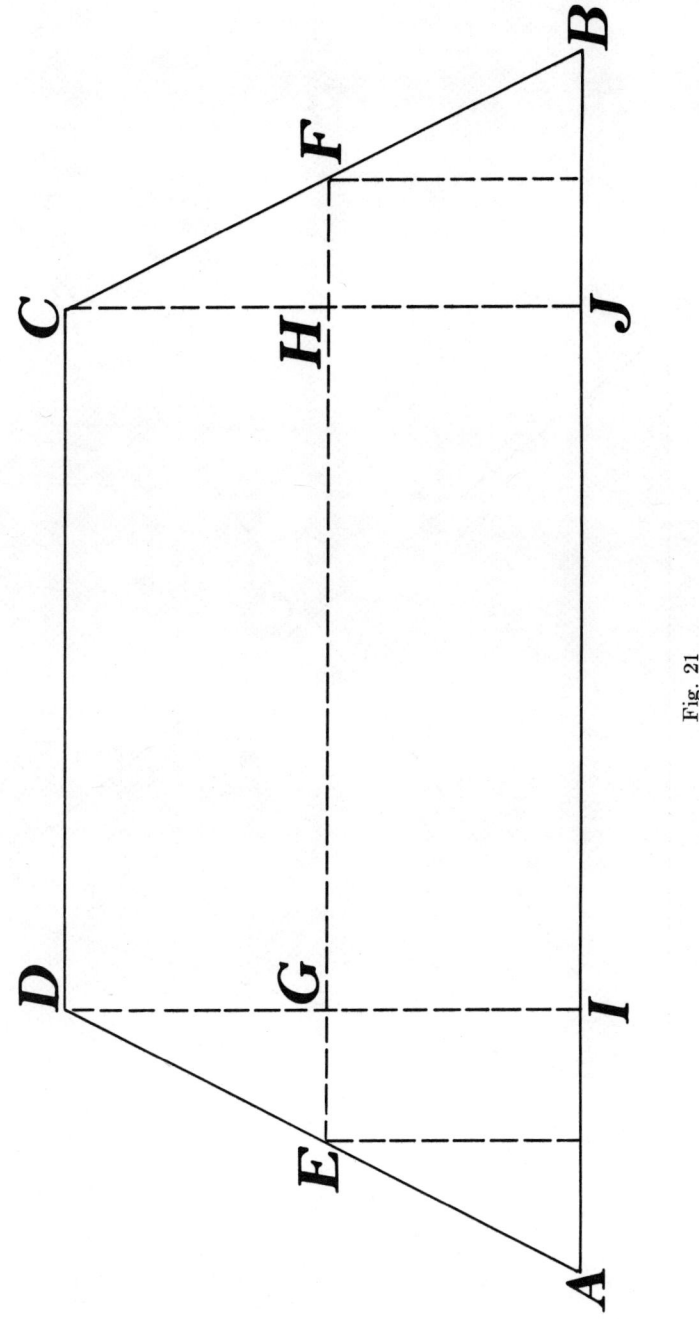

Fig. 21

51

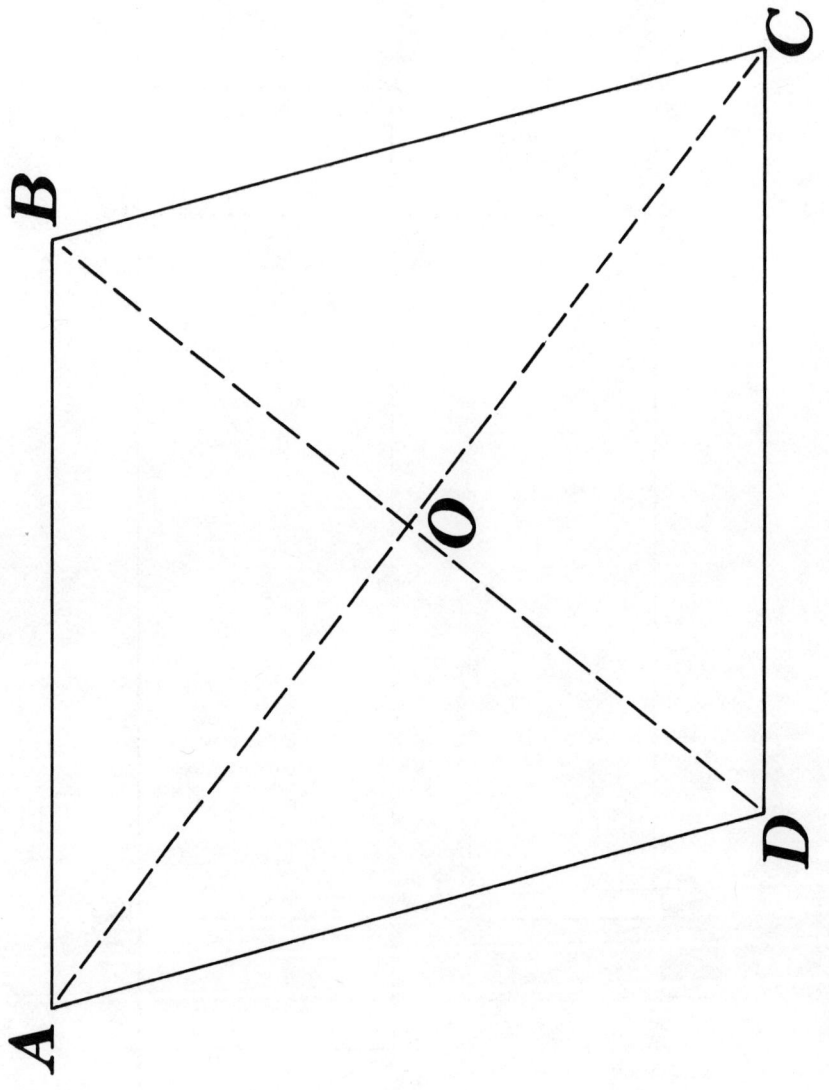

Fig. 22

52

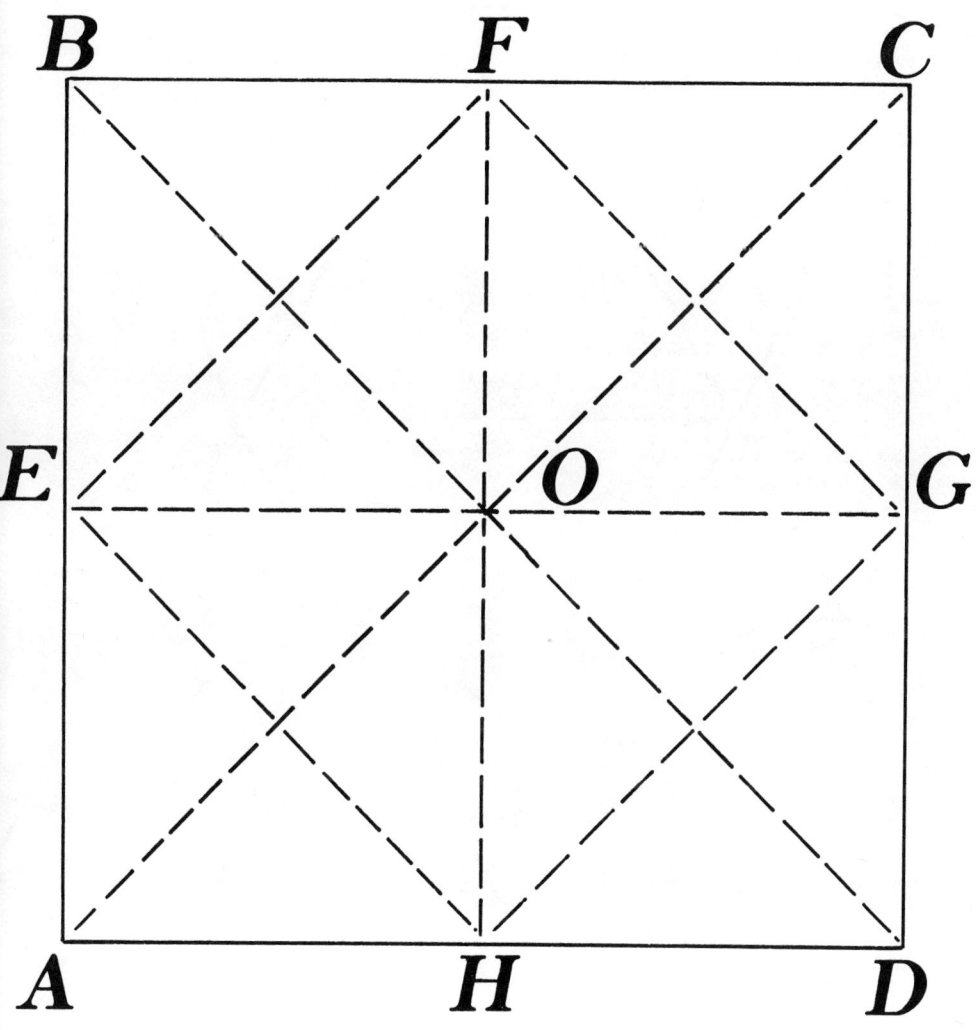

Fig. 47A

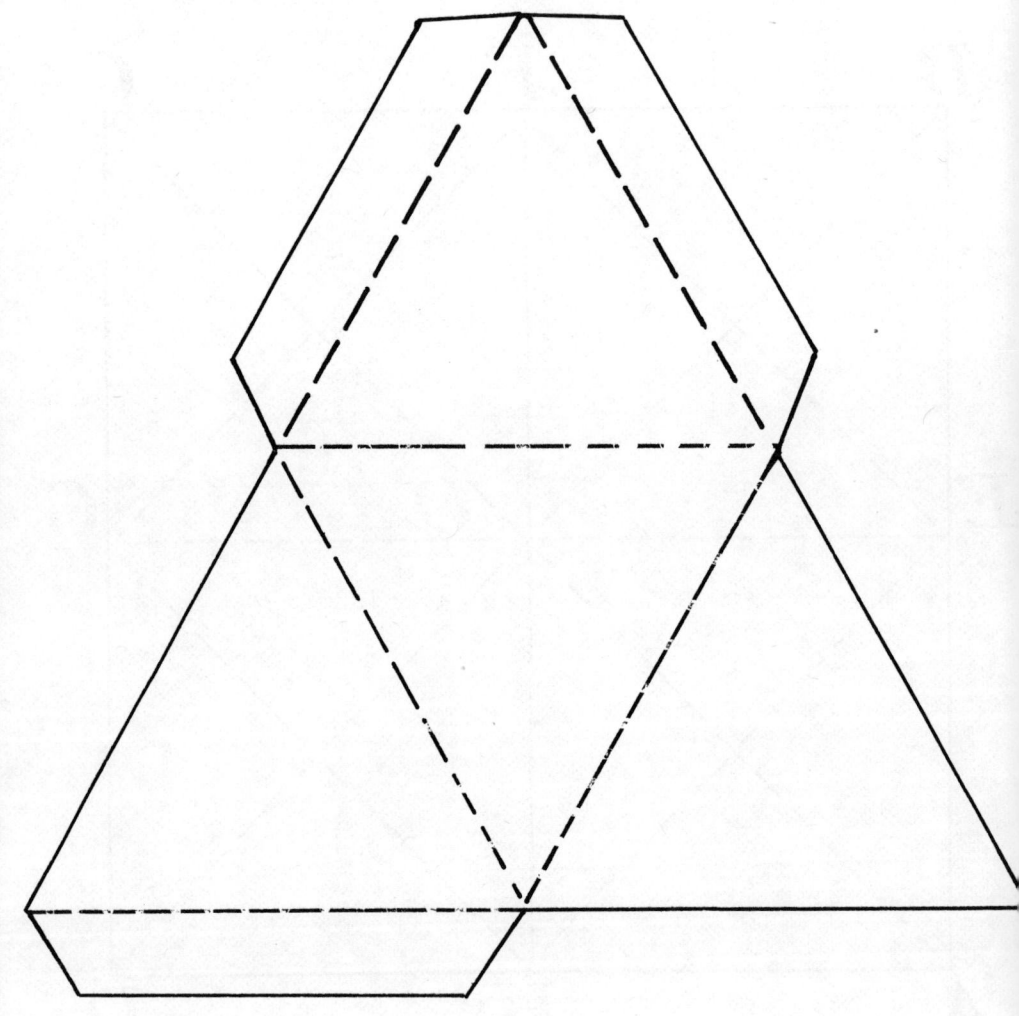

Fig. 72A

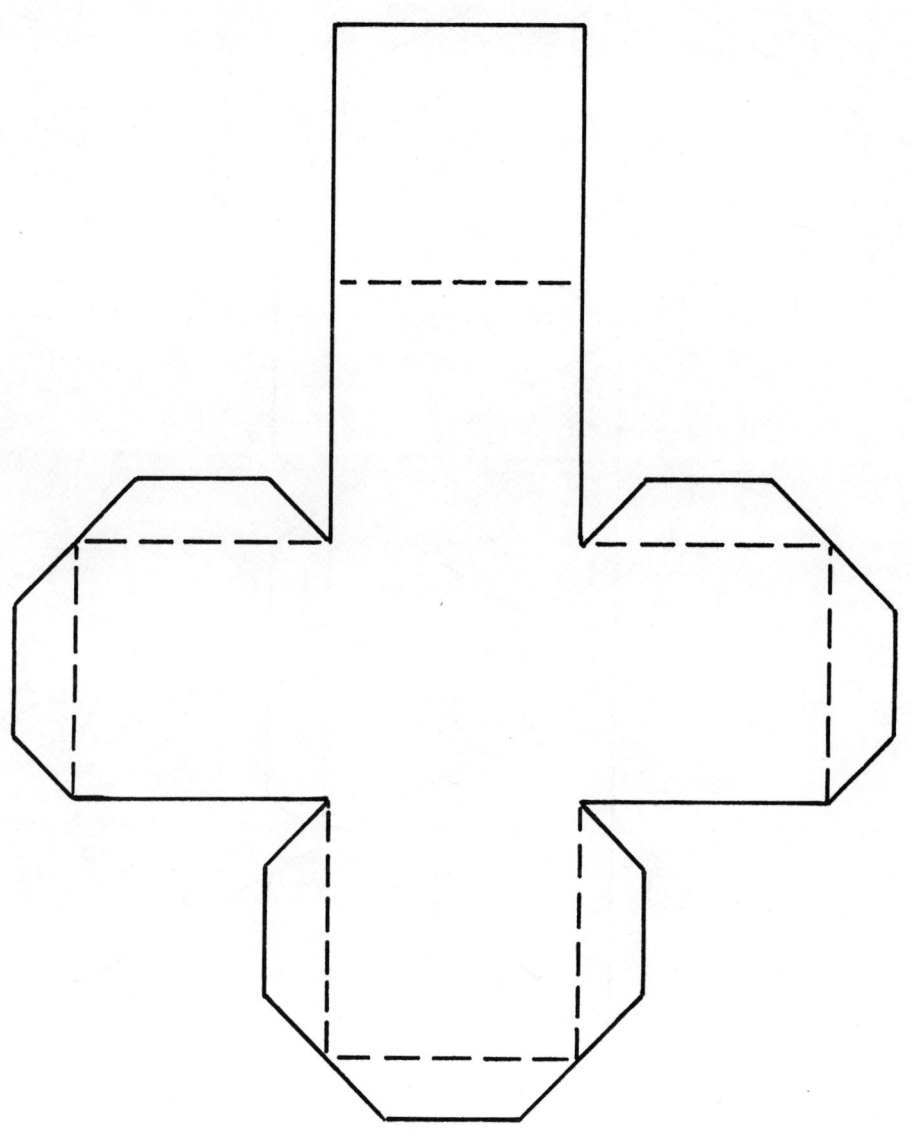

Fig. 72B

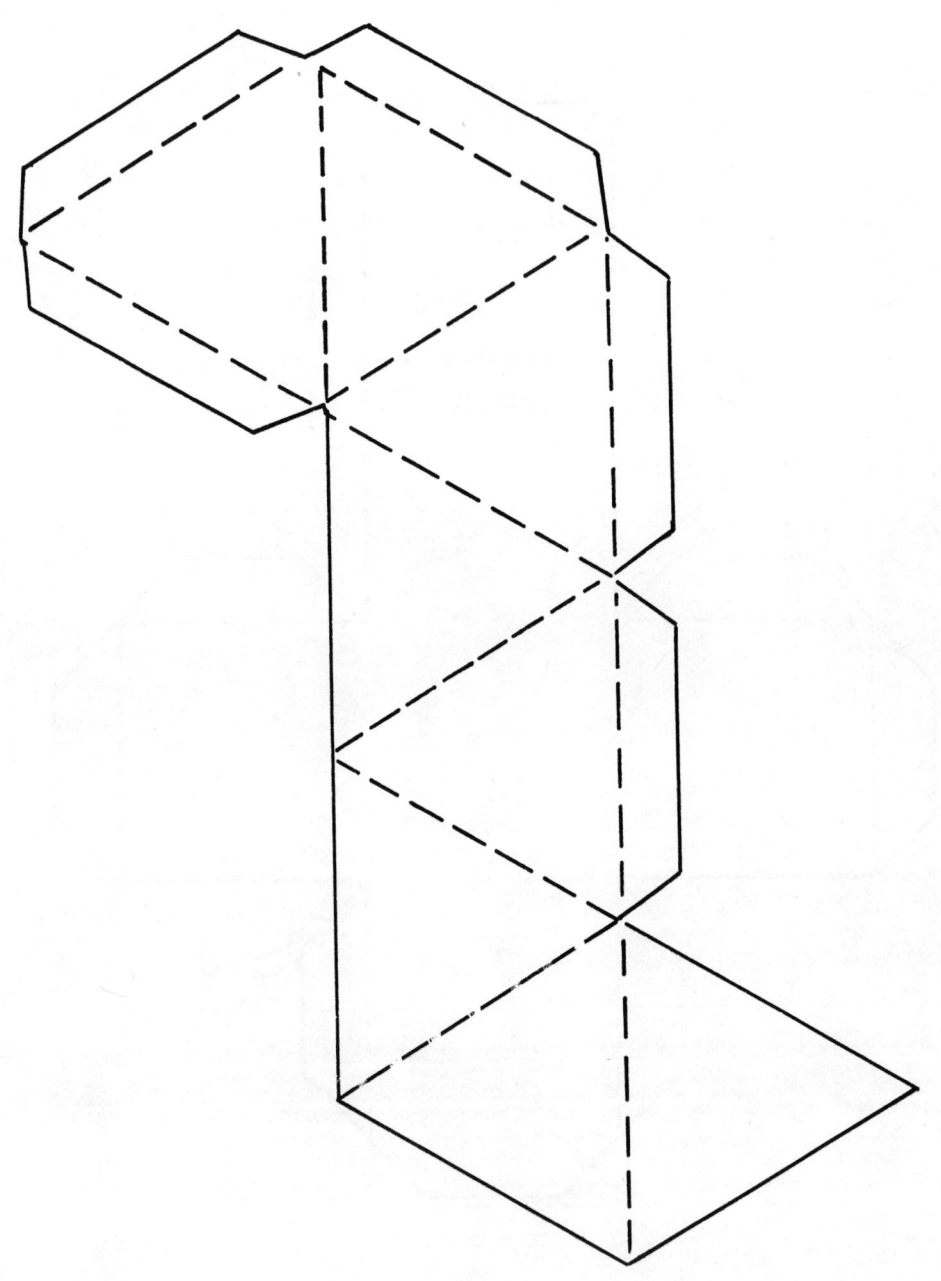

Fig. 72C

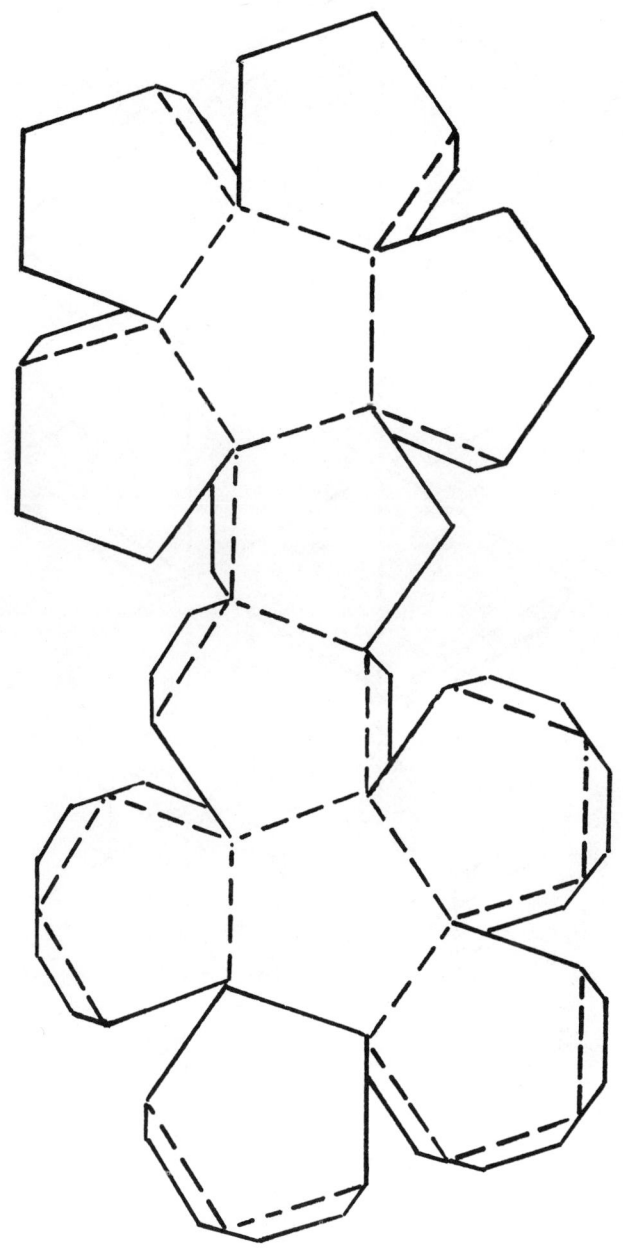

Fig. 72D

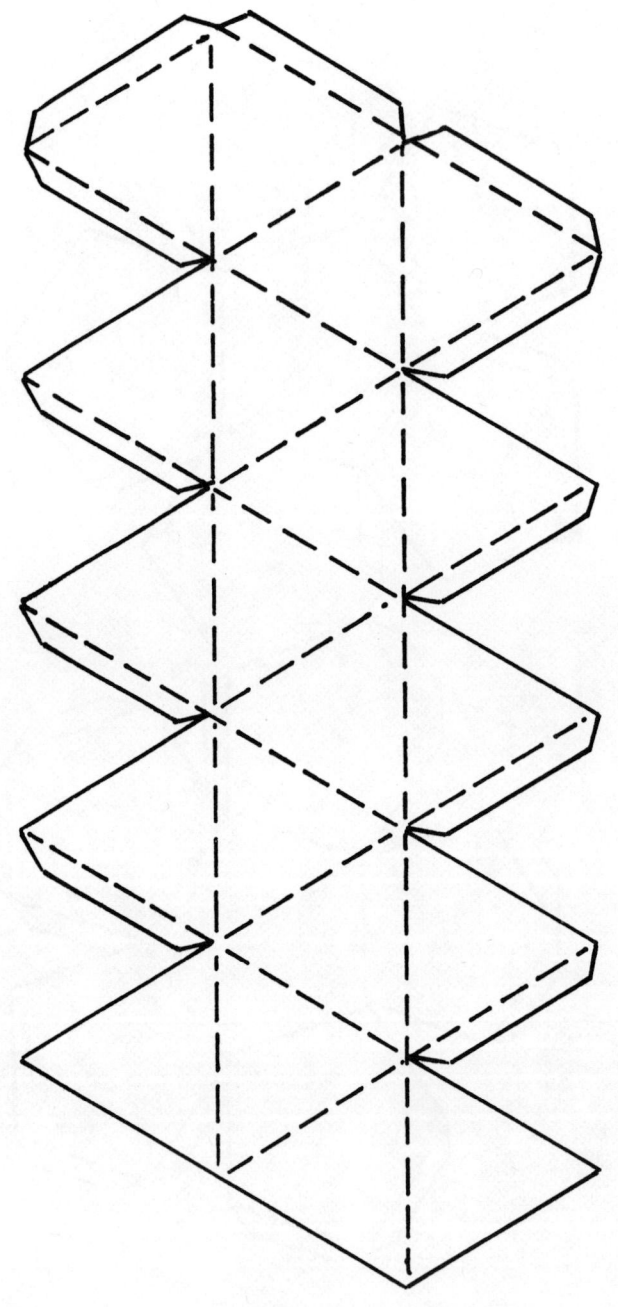

Fig. 72E

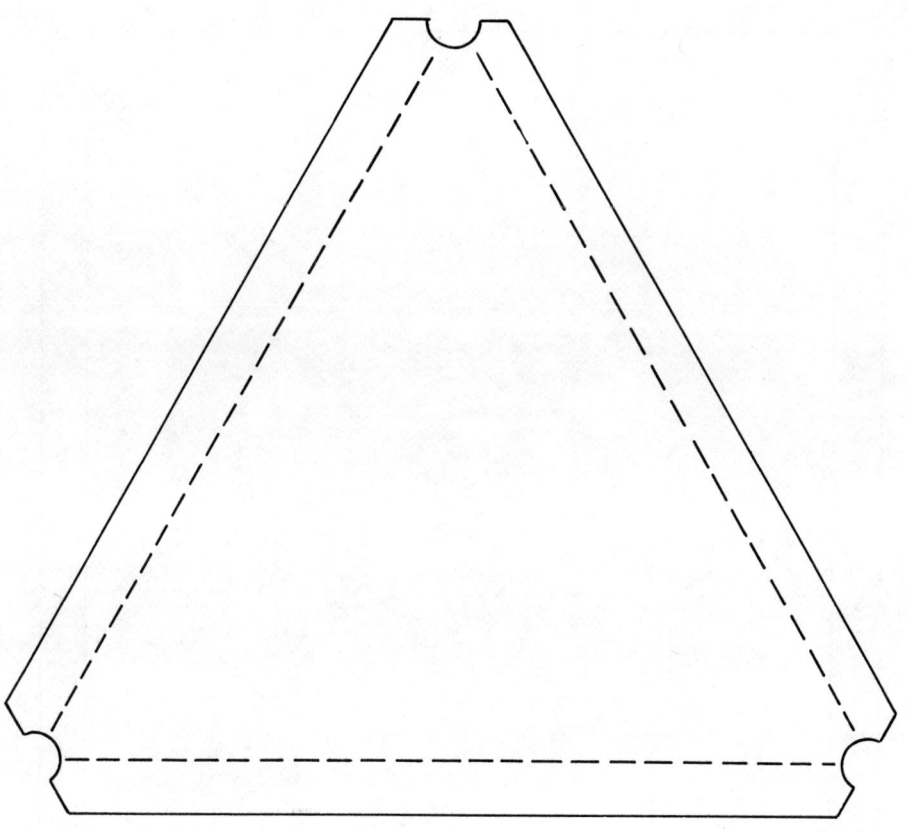

Fig. 72F

Fig. 72G